LEEDS CO**LLEGE OF BUILDING LIBRARY**

CLASS NO

BARCODE

✔ KT-450-034

FLOORING

INSTANT ANSWERS

LEEDS COLLEGE OF BUILDING
WITHDRAWN FROM STOCK

FLOORING INSTANT ANSWERS

Steven J. Bukowski

McGRAW-HILL

New York   Chicago   San Francisco   Lisbon   London
Madrid   Mexico City   Milan   New Delhi   San Juan
Seoul   Singapore   Sydney   Toronto

The McGraw·Hill Companies

Copyright © 2003 by The McGraw-Hill Companies, Inc. All rights reserved. Printed in the United States of America. Except as permitted under the United States Copyright Act of 1976, no part of this publication may be reproduced or distributed in any form or by any means, or stored in a data base or retrieval system, without the prior written permission of the publisher.

1 2 3 4 5 6 7 8 9 0 DOC/DOC 0 9 8 7 6 5 4 3 2

ISBN 0-07-140204-7   .

The sponsoring editor for this book was Larry S. Hager and the production supervisor was Sherri Souffrance. It was set in Stone Sans by Lone Wolf Enterprises, Ltd.

Printed and bound by RR Donnelley.

This book is printed on recycled, acid-free paper containing a minimum of 50% recycled, de-inked fiber.

McGraw-Hill books are available at special quantity discounts to use as premiums and sales promotions, or for use in corporate training programs. For more information, please write to the Director of Special Sales, McGraw-Hill Professional, Two Penn Plaza, New York, NY 10121-2298. Or contact your local bookstore.

Information contained in this work has been obtained by The McGraw-Hill Companies, Inc. ("McGraw-Hill") from sources believed to be reliable. However, neither McGraw-Hill nor its authors guarantee the accuracy or completeness of any information published herein, and neither McGraw-Hill nor its authors shall be responsible for any errors, omissions, or damages arising out of use of this information. This work is published with the understanding that McGraw-Hill and its authors are supplying information but are not attempting to render engineering or other professional services. If such services are required, the assistance of an appropriate professional should be sought.

CONTENTS

ACKNOWLEDGMENTS

A & D Tile Mart
8320 S. Pulaski Road
Chicago, IL 60652
(773) 581-7580

Armstrong Worldwide Industries, Inc.
2500 Columbia Avenue
Lancaster, PA 17604
(800) 233-3823

Bestile, Inc.
15555 S. 71st Court
Orland Park, IL 60462
(708) 532-6006

CarpetMax
9930 W. 55th
Countryside, IL
(708) 352-8300

Chicago Antique Brick
2201 S. Halsted
Chicago, IL 60608
(312) 666-3257

Century Tile
20909 S. Cicero Avenue
Matteson, IL 60433
(708) 503-4700

Crain Tools
1155 Wrigley Way
Milpitas, CA 95035
(409) 946-6100

Edelman Leather
80 Pickett District Drive
New Milford, CT 06676
(860) 350-9600

Flooring Network
1366 Merchandise Mart
Chicago, IL 60654
(312) 321-1217

Hoboken Floors
70 Demarest Drive
Wayne, NJ 07470
(973) 694-2888

Hart to Hart Interiors, Inc.
P.O. Box 104
Flossmoor, IL 60422
(708) 214-1951

Illinois Brick
1300 W. Maple
Mokena, IL
(815) 485-2533

Mannington Wood Floors
1327 Lincoln Drive
High Point, NC 27260
(336) 884-5600

Midwest Padding
2500 Old Hadar Road
P.O. Box 2283
Norfolk, NE 68702
(402) 379 9737

Mintec Corp.
100 E. Pennsylvania Avenue
Towson, MD 21286
(888) 964 6832

Natural Cork
1710 North Leg Court
Augusta, GA 30909
706-733-6120

P.D. Hartz Construction Co., Inc.
8995 W. 95th Street
Palos Hills, IL 60465
708-233-3800

Pergo Flooring
P.O. Box 1775
Horsham, PA 19044
(800) 337-3746

Plyboo Flooring
375 Oster Point Boulevard #3
S. San Francisco, CA 94080
(650) 872-1184

Hanson Real Brick
3820 Serr Road
Corunna, MI 48817
(800) 447-7440

Universal Flooring
2009 Chenault Drive
Carrollton, TX 75006
(972) 387-9315

ABOUT THE AUTHOR

Steven Bukowski has a degree in interior design and runs an interior design firm which specializes in model home design. He has been in the construction and design industry since 1985.

INTRODUCTION

This book is about the installation of flooring, covering a wide variety of types: from hardwood to marble, carpet to resilient vinyl, and other combinations you may not have thought about. You will see how to install flooring, what tools are needed, and a number of tricks of the trade necessary to install the floor in an organized and professional way. Before getting to the actual installation of each floor, let's briefly explain a little about each type of flooring.

RESILIENT VINYL FLOORING

Resilient vinyl flooring is a flexible product that can come on large rolls 6 to 12 feet wide. It is usually glued to an underlayment or on self-adhesive backed tiles of 12 inches square. This product is available in many types of patterns and colors and usually has a slight texture on the surface.

CERAMIC TILE

Traditional ceramic tiles are usually made from oven-baked clay, while natural stone tiles are quarried and cut from stone around the world, the highest quality being from Italy. Ceramic is a very durable material and available in a large array of colors, shapes and sizes. Other types of natural tile are made from slate, granite and marble.

HARDWOOD FLOORING

Hardwood is available in either solid or veneered wood and laminated planks, as well as parquet squares. The most common woods used are maple and oak. Most hardwood planks are manufactured with a tongue and groove construction, so that when they are placed side by side, they lock together with a minimal gap between them. Laminate wood planks are usually backed with compressed wood and glued to the actual laminate wood veneer.

Resilient vinyl tile is the least expensive choice in floor covering; it is attractive and durable, and commonly used in kitchens and baths.

CARPET

Carpet is made of soft, flexible fibers bonded on the back by a mesh material. It is usually manufactured in rolls 12 feet wide and can be made wider by seaming it together. Available in a huge array of solid colors, it is also an option to order custom colors or patterns to match a rooms decor.

There are two types of carpet: loop pile and cut-pile. The loop-pile creates a textured look and feel, while the cut-pile makes the carpet design look very consistent and solid.

When you come to the realization that you are ready for a new floor, the list of things to access can help in determining the likenesses and differences in the variety of flooring available in today's market. The list is short, but concise.

Ceramic tile is a great choice for bathrooms and kitchens as it offers easy maintenance and a simple, clean look.

Marble tile has a high-end look that is often well worth the extra cost. This tile works best in a formal setting.

Hardwood flooring is made of oak or maple. This type of floor can add warmth and create a comfortable feeling.

Low-pile carpet can can be used for a contemporary or formal look, and is great for sound absorption.

Flooring selections in today's market are widely varied in purpose, price, and quality. Several factors help dictate choices in new flooring:

- *Safety*
- *Wear and tear*
- *Comfort*
- *Noise level*
- *Maintenance*
- *Cost and amount of labor involved*

SAFETY

The issue of safety comes into play when considering where the flooring is going to be located. You want to put the proper product where it will be most useful in creating a safe place to work or play. For example, you would most likely want to place a resilient sheet vinyl in a laundry room rather than carpet, because there is a good possibility that water will gather on the floor from time to time. Vinyl is easier to clean up, and more resistant to short term water spills. Carpet would be a poor choice because it will absorb the water and mold can collect between the layers of padding and carpet, unless it is dried thoroughly every time it gets wet.

Using marble on a kitchen or bathroom floor can create dangerous slipping risks when wet. A better option would be to use resilient vinyl flooring, which has a slight texture that is designed into the surface for better traction.

WEAR AND TEAR

When it comes to wear and tear, the choices of flooring materials should match the intended use. Whether it's in a large, busy household or an office with steady streams of foot traffic, beautiful results can be achieved while still getting the most wear from the flooring products chosen. Carpet is the most popular flooring surface in most most family rooms. Carpet should be installed with a minimum of 6 lb. weight pad. This improves the wearability of the carpet by creating a buffer between the carpet and sub-floor.

While there is a trend toward moving to hardwood floors, it is far easier to maintain a carpeted floor when it comes to dirt and scratches from shoes and furniture. Most carpet already comes with stain resistance built into the fibers, and can be vacuumed once a week and cleaned every six months. When a hardwood floor is scratched or becomes dirty, the surface of the floor tends to reflect some element of damage even after it has been cleaned or repaired. Most new laminate floors still scratch easily and have to be swept or mopped daily for optimum life.

COMFORT

The comfort and aesthetic qualities of a room are probably the most important reasons a new floor is being installed, so this choice should be made carefully. The color of the floor is also a part of the comfort factor. Take a look at the existing furniture and wall color in the room. If any of those items are being changed after the new floor is installed, get a swatch of the new wall and furniture fabrics and colors. Make sure it will be a harmonious match, since the floor will be around for a long time and will be difficult and expensive to change.

If a theme is involved, choose flooring that will lends itself to the theme. For example, if an old-world theme is being used in a kitchen, a good option is to use slate flooring that mixes and matches with cabinets and wall colors.

NOISE LEVEL

Carpet is a good choice for noise abatement, and can make a room appear soft and quiet. Resilient vinyl is much quieter than ceramic tile for foot traffic, and is also easier to maintain. It is a much quieter choice when it comes to walking over it with shoes and is easier to maintain.

COST AND LABOR

The costs of material and labor are often the deciding factors in a homeowner's flooring decisions. There are a number of options in flooring methods and choices that can help reduce material costs, while maintaining a high level of quality.

An expensive berber carpet that may be just out of a homeowner's price range can be made much more affordable by installing resilient tile in the foyer and closets, saving the more expensive material for the main floors.

Ceramic tile in bathrooms is a terrific choice, but can also be an expensive one for the homeowner. Options can be using resilient tiling that mimics the look of ceramic in the guest bathroom, while reserving the ceramic tiling for the master bath.

MAINTENANCE

Resilient tile floors are the easiest to maintain. They can be damp mopped on a weekly basis or simply swept on occasion. If they are used in a kitchen, they can be a great barrier for trapping mud and dirt before it gets to the carpet in other rooms.

Carpets are becoming easier to maintain because of modern stain resistant fibers, either already built-in or applied after installation. Vacuuming on a weekly basis and cleaning every six months can help to maintain the longevity of a carpet. Ceramic tile floors are somewhat more difficult to maintain in preserving their original luster. Texture and color can sometimes hide the discoloration of grout, but dirt still becomes embedded, and it takes a lot of scrubbing with cleaning solutions to keep it clean.

Hardwood flooring is easy to maintain in regards to cleanliness, however hardwood will become scratched under use from appliances, shoes, pets, and the hazards of everyday living. A damp mop on a weekly basis keeps it clean, but the repair of scratches with a wood fill pen or liquid oil can become more trouble than its worth.

When it comes to original hardwood flooring, even more maintenance is needed. The use of furniture oil, and in extreme situations, complete stripping and replacement of varnish is needed to keep the floor in good shape and protect it from further damage.

INSTALLATION

At the beginning of each flooring type installation in the book we will show how to measure and calculate accurately so there is no shortage of materials once the project is underway. We will also list the tools needed for each type of floor installation.

When a floor is installed professionally and maintained properly, the results will last, and will add tremendously to the quality of everyday living for the homeowner.

CARPET INSTALLATION

Wall-to-wall carpeting is one of the easiest and quickest ways to improve the look of most floors. Carpet adds warmth and color and comfort to any room. Most people want the floor to match the furnishings in the room. Since carpet has the most variety of color and texture, it becomes an obvious choice. Along with variety and color, carpet can be installed for a reasonable cost. All of these factors play a part in the decision of choosing carpet as the flooring of choice for the project.

CARPET CONSTRUCTION

Let's look at some information on the construction of most types of carpet. All carpets are made of fibers; the five main types are nylon, polyester, olefin, acrylic and wool.

Nylon fibers are easy to clean, very durable, and offer good stain resistance. Polyester fibers offer greater stain resistance than nylon fibers and the colors do not tend to fade in sunlight after a long period of time. Olefin fibers are virtually stain and moisture proof, but they do not stand up to heavy use and are not as soft as nylon or polyester fibers. Acrylic fibers are soft in feel and look, generally moisture resistant, but less durable than the other fibers.

Wool fibers are the most expensive of all and offer good durability and warmth (Figure 1.1).

PROPERTY	FIBER						
	WOOL	NYLON	NYLON 6-6	ACRYLIC	MODACRYLIC	OLEFIN	POLYESTER
ABRASION RESISTANCE	good	excellent	excellent	good	good	excellent	good
RESILIENCY	very good	excellent	very good	very good	very good	good	fair
COLOR RETENTION	fair	very good	very good	good	good	excellent	good
PATTERN RETENTION	good	very good	very good	good	good	good	fair
TEXTURE RETENTION	good	very good	very good	good	good	good	fair
STAIN AND SOIL RESISTANCE	good	good	excellent	good	good	excellent	good
STAIN AND SOIL REMOVAL	good	good	very good	good	good	excellent	good
HEAT/FLAME RESISTANCE	excellent	very good	very good	good*	very good	good	good
MOISTURE ABSORBENCY	high	low	low	low	low	very low	low
MOLD/MILDEW, INSECT RESISTANCE	†	high	high	high	high	high	high
STATIC RESISTANCE	low	low	high	medium	medium	excellent	low

†requires treatment   *if treated

FIGURE 1.1   Carpet fiber property comparison chart.

One way to find out if you are looking at a good quality carpet is to examine the pile density. The denser the pile, the more it will hold up against traffic and stains. If you look on the backside of the carpet, the tighter the grid pattern is, the denser the pile will be.

TYPES OF CARPETING

There are four basic types of carpet available on the market today. Cushion-backed carpet eliminates the need for extra padding because it already has foam attached to the back. This type of carpet is easier to install than others because there is no need to stretch it or lay down tacking strips. It is simply secured by a general-purpose adhesive. Cushion-back carpet is usually of a lesser quality, therefore it costs less. Remember, your customers get what they pay for and may end up replacing the carpet, adding to the long term cost of a higher quality carpet they really like that will last much longer.

Another type of carpet is loop-pile carpet, which simply means what it says, a carpet made completely of a series of loops which gives it a great textured look (Figure 1.2). The loops can be arranged in a specific pattern or randomly. This type of carpet is ideal for heavy traffic areas. In the marketplace today, a type of loop-pile carpet known as *berber* is the most popular style requested for new home construction.

The third general carpet type is velvet-cut pile (Figure 1.3). This has the most uniform appearance and thickest pile of any carpet. It has a very traditional look and works well in a living room or dining setting.

Saxony cut-pile, commonly called *plush* (Figure 1.4), is great in a family room setting. This type of carpet holds up against traffic and keeps more dirt from building up in the fibers over a long period of time. Its appearance is speckled, which can help hide light dirt and wear.

CARPET LABELS

The best way to learn about carpet is to flip over carpet swatches and read the labels on the back (Figure 1.5). The label should give you the type of fibers used to manufacture the carpet, whether it has any additional stain resistance chemicals added to it (which may break down the strength of the fiber but give you better stain resistance),

FIGURE 1.2 Loop pile carpet.

and what dimensions it is available in, which will come in handy when figuring out room sizes.

Also, the label may show additional performance features and what the manufacturer's warranty will cover and for how long. The better the warranty, the better quality the carpet usually is.

TYPES OF PADDING

There are several types of padding you can use, depending on the type of carpet, the environment you are putting it into, traffic conditions, noise reduction, insulation value, and cost. Padding can be categorized into three groups:

FIGURE 1.3   Velvet-cut pile.

- *Fibrous cushions*
- *Cellular sponge or rubber*
- *Urethane foam*

Fibrous Padding

Fibrous cushion provides a firm, dense cushion preferred for heavy, medium or light traffic flow (Figure 1.6). This type of padding has a tendency to attract moisture, so a low-grade moisture barrier, such as a latex sheet, should be placed between the padding and the carpet.

When selecting this felt-type padding, you can also have the option of placing a rubber backing on it. This stops slippage and improves the longevity of the carpet.

FIGURE 1.4 Plush carpet.

Cellular Sponge or Rubber Padding

Another choice for padding is sponge or foam rubber (Figure 1.7). While the sponge type of cushioning is great for that plush feeling, it is not recommended for heavy traffic areas. The foam rubber padding is a firmer, tighter padding which can hold up to heavier traffic. Foam is a product which can retain odors if spilled on, so you may want to be careful where you place this type of padding.

Urethane Padding

Urethane padding is suitable for use in light to heavy traffic. Over time, however, this type of cushioning may bottom out or flatten, so it is important to get the highest density to lessen that possibility. Do not use fillers or more than one percent of waste material in your padding.

FIGURE 1.5 Typical carpet label.

7

FIGURE 1.6   Fibrous cushion.

All types of padding should be installed in the largest possible lengths, with a minimum of seaming. Also, when the carpet is placed over the padding, the padding seams should not be the same as the carpet seams. The installation of padding is even more important than the actual carpet. Be sure the padding is flat to insure safety, keeping the carpet from becoming rippled or showing seams. Choosing the right type of padding can increase the longevity and improve the feel of the carpet. One thing to keep in mind, choosing a thicker pad does not improve the quality of the carpet; it actually decreases the wearability by causing the padding to rub against the carpet and literally wear it out. The thicker pad has a heavier volume than the thinner carpet and, after a short amount of time, natural stretching occurs from walking over the carpet. This stretching causes the carpet to rub against the padding and wear it out faster than it would with the proper weight of padding for the carpet.

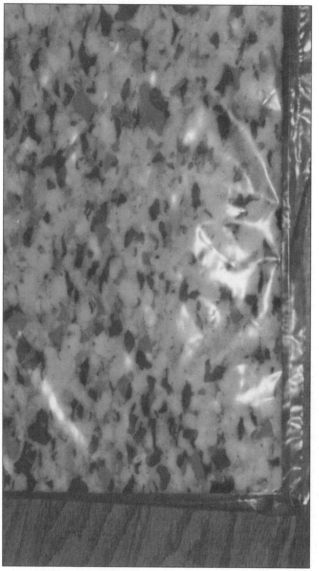

FIGURE 1.7 Sponge padding.

MEASURING FOR PADDING AND CARPET

The same measurements and calculations are used for padding and carpet. Multiply the width by the length to find the total area of a room. Any extra bays or areas not within that field should be measured separately and added to the total floor area measurement. Most carpet is made in widths of 12 or 15 feet.

When you are measuring a room and want to make a specific pattern work, keep those widths in mind. If the room is wider, you will have to have the carpet seamed which may throw off the pattern in your carpet. If it is a solid or speckled carpet, it is much easier to hide the seam. Over a period of months, the seam should be worn into the carpet after normal use and your customer will hardly know where it is. You should not place a seam where there will be a heavy traffic pattern. If a seam is necessary, be sure to add an additional 20 percent to the total area to insure proper cutting and finishing. Adding 20 percent to your product measurement will insure that you have enough carpet to properly seam and trim the area.

Measuring Closets and Stairways

If you are measuring a room with closets, do not take the full piece of carpet from the total area and try to cut the carpet for the closets. Cut it separately from the same piece and seam it into those areas. When measuring for padding, you do not need to add an additional 20 percent to your calculation. The padding will not need to be stretched or seamed like the carpet.

To measure stairs for carpet, add the rise and run, also known as the front and top, of each stair. Then measure the width of the stairway to find out how many cuts you can get from your roll of carpet. Rather than seaming all the pieces together, attach each strip so it ends at a seam in the step or under the runner.

TOOLS NEEDED FOR CARPET INSTALLATION

Once you have properly measured the entire area you are installing, you are ready to gather the proper tools for the job. What follows is a checklist of tools you will need.

Tool Checklist

✔ Power stretcher and extensions. Make sure your extensions are the width of the room.

✔ Seam iron

✔ Knee kicker

✔ Chalk line

✔ Edge trimmer

✔ Row-running knife

✔ Regular utility knife

✔ Stair tool

✔ Hammer

✔ Snip, scissors and stapler

Additional Supplies for Installing Padding

For the padding installation, you will need some additional supplies. You may need hot-glue seam tape to join the carpet seams together. Duct tape will be used to join padding seams together. Double-sided tape can be used to attach carpet to concrete floors with staples used to connect padding to wooden underlayment.

Tackless strips, which are small strips of wood with little tacks sticking up through it, are used on the edges of the sub-floor to hold on to edges of stretched carpet. These strips are 1½ inches wide with rust resistant metal pins pointing upwards. Three rows of pins are recommended for heavy traffic, while two rows is sufficient for all other areas. You need to place these strips around the perimeter of the room leaving a small gap between them and the wall.

SUB-FLOORS

Sub-floors must be properly prepared and cleaned before laying down the pad and carpet. Sub-floors can be plywood, wood plank flooring, or even cement. No matter what they are, it is important to get them as level and smooth as possible. Also, when preparing the sub-floors, be sure to remove all transitions between rooms or different types of flooring (Figure 1.8).

Transitions are necessary when there are two different types of floors. They finish or cover the area where the flooring comes together. A reducer strip finishes the space between a laminate floor and carpet. The transition can be applied using screws, nails or adhesives. Any edges of carpet not protected by a transition are likely to turn up over time and cause a safety issue.

Concrete Sub-floor

Concrete sub-floors will likely accept all types of carpet. The way to install carpet on them is by using an adhesive. If the concrete floor is below grade, such as a basement application, it should be treated with moisture blocker before you apply the adhesive. Make sure the floor is swept clean and fill in any tiny holes, larger than ⅛ inch in

FIGURE 1.8 Transition between tile and carpeted area.

diameter, with floor leveling compound. If the concrete is severely cracked and uneven, you may want to apply an entirely new surface of leveling compound. This will level the floor and cover all gaps and imperfections. Leveling compound is available in five gallon buckets. Read the manufacturers' directions carefully before pouring it onto the surface. This type of mixture usually levels itself out, so not much effort is necessary. Just smooth it out as needed and let dry for 24 hours before applying the moisture barrier or new sub-floor. There are generally four types of adhesives you can use:

- *Natural latex*
- *Synthetic latex*
- *Solvent-based adhesive*
- *Alcohol solvent resins*

In most cases, the natural and synthetic latex products are suitable and safest. Always read the manufacturers' instructions carefully before using their product. If you do not use an adhesive to install the padding to your concrete floor, you may use the tackless strips by drilling them directly into your concrete sub-floor.

Plywood Sub-floors

Plywood sub-floors are needed when there are too many gaps to fill in the wood floor or if you are going to cover two or more different types of flooring. When using plywood for a sub-floor, a sturdy ¾ inch stock sheet should be used. When nailing it down, nail around the entire perimeter of the piece of plywood using annular nails. Make sure to stagger the corners of each piece of plywood so they never meet (Figure 1.9). Over time, common corners could raise or buckle under pressure, especially in heavy traffic areas. Be sure to cover the entire area being carpeted with the sub-floor. When all is nailed down, go over the floor surface to make sure no nail heads are sticking up and that all edges of the plywood are flush with each other. If the nail heads have caused a hole larger than ⅛ inch, fill them in with wood filler. It is important that the floor be as smooth as possible. Once the carpet is installed, over time these untreated imperfections will be felt under the carpet.

Once the sub-floor is smooth, clean, and level, you are ready to apply the tackless strips. You should wear a pair of work gloves for this part of the installation—it could get a little prickly. Take the strips, facing the angle of the pins toward the wall, and place

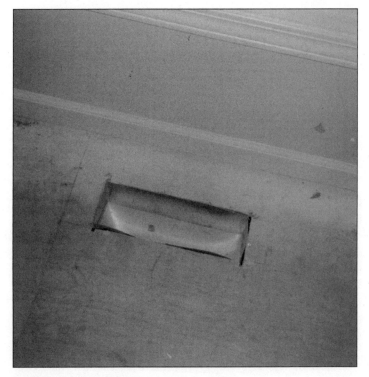

FIGURE 1.9 Staggered corners on plywood floor.

around the entire perimeter of the room (Figure 1.10). Leave a small gap between the wall and the strip, usually equal to ⅔ the thickness of the carpet. Use a piece of cardboard to space your strips evenly and to protect the wall as you work around the room. This allows space to tuck the edges of carpet into and reinstall any trim work after installation of the carpet. To install the strips, you can nail them directly into the wood sub-floor, closely spacing the nails an inch or two apart in order to hold the strips securely. You may also drill screws into the strips to hold them down. This type of installation is usually only necessary for concrete floor installations.

FIGURE 1.10 Tackless strips applied to plywood sub-floor.

fastfacts

Be careful when installing tackless strips into new or existing floors. There may be radiant heating or other pipes which could puncture when a nail or screw are driven through them. A good way to find out where radiant heating might be located is to spray water on the edge of the wall where the tackless strips are going to be located. Turn up the heat and mark the areas that dry first, this is where the radiant heating is situated.

fastfacts

Before laying the plywood sub-floor, the boards should be conditioned. Apply a thin coat of water on one side then lay them flat on top of one another in the room they are to used for at least 48 hours. By doing this process it allows the wood to acclimate to the room temperature and prevent warping in the future.

INSTALLATION OF PADDING

Installation of the padding depends on where the flooring is located and on what type of sub-floor is it being applied. You can apply padding to a concrete or plywood sub-floor.

Installation of Padding on a Concrete Floor

If the flooring is being located in a basement area, or a below grade area you will probably be using an adhesive to install your padding. This method is commonly used in these types of areas and should secure the padding. Once the area is smooth, level and clean, spread the adhesive in an area you can work with fairly quickly. A thin layer is sufficient as long as the entire area is completely covered. Unroll the padding and press firmly until it is even. Go around the perimeter and trim off the excess padding with a utility knife. If you are using the tackless strip method for installing the carpet, cut the padding so the strips are left uncovered.

If the roll is not large enough to cover the entire area, you may need to seam pieces together. Any uncovered areas should be measured and have pieces of padding cut to fit. These smaller pieces should be taped together at all the seams with duct tape to make it one overall piece. If you do not use the adhesive for applying the padding, you may also use double-sided tape.

Installation of Padding on a Plywood Sub-floor

If you are installing padding in a grade or above grade room, such as a bedroom, you are most likely applying padding to a plywood sub-floor. Unroll the padding and cut to fit the entire area to be carpeted (Figure 1.11). Once you have covered your area, cut away all excess padding, leaving the tackless strips exposed. Tape the seams, if necessary, with duct tape and staple down the padding directly to the plywood sub-floor about every 6 inches (Figure 1.12). This should secure it tightly so that it will not bunch under the carpet.

CARPET INSTALLATION

We are going to cover two types of carpet installation. Both have many similarities, but also have some important differences. Cushion-backed carpet installation and wall-to-wall carpet installation are what most people use today.

FIGURE 1.11 Adding padding to a plywood sub-floor.

FIGURE 1.12 Taping the seams of padding.

Cushion-Backed Carpet Installation

Cushion-backed carpet is installed much the same way as wall-to-wall, except it is not stretched and does not require tackless strips. A full bond adhesive is used to attach this type of carpet to a floor. The cushion is already applied to the back of this carpet so it does not need an additional layer of padding. You can then apply it directly to your sub-floor with the adhesive. This type of carpet and installation is the least expensive and is usually applied to lower quality carpet.

After you've measured and cut your carpet, simply spread the adhesive and unroll the carpet, pressing down firmly as you unroll. Cut away any excess carpet and tuck away your edges using a putty knife.

fastfacts

The best way to carry your carpet from the installation area to the cutting area is to fold both long sides toward the center, then loosely roll it up. The roll is bulkier, but easier to maneuver through doorways and around corners. Always roll and unroll the carpet in the direction of the pile.

Installing Wall-to-Wall Carpet

Wall-to-wall carpeting installation is a matter of stretching and making sure the seams match and are not noticeable. The process of seaming can be tricky and should be practiced on a scrap piece of carpet before tackling your installation. Go ahead and measure the area to be covered. You can fold the carpet in half and carry it into the room being covered. Lay out the carpet and mark it off on the back using a dark marker or pencil, leaving an excess of about 3 to 6 inches around the perimeter of the wall to allow for stretching and seaming. Take the carpet out of the room and place in a large area, such as a clean garage floor or driveway, and cut with the utility knife according to the marks on back of the carpet. Bring the cut piece back into the room and place it as close as possible on top of tackless strips secured around the perimeter of room. Go around and flatten out the carpet as much as possible.

On each corner, cut into the carpet at a diagonal to help the carpet lay flat. Also, cut any slits you need to help place the carpet around radiator pipes or other obstructions.

If you are using a piece of carpet which does not cover the entire area of the room, you will need to make a seam somewhere in the carpet. This should be placed in an area where there is not a lot of traffic and is not highly visible. You should also run a seam with the direction of the source of light; this helps minimize the shadow which may highlight the seam. Where the two pieces of carpet will meet at the new seam, pull back the carpet and make a line about 2 inches away from the edge. Now get the seam iron plugged in and let it heat up.

Cut a piece of seam tape the length of the seam and place it under the carpet, centering it under the new seam. Place your seam iron under the carpet and directly on the tape and let the tape liquefy for

fastfacts

When cutting carpet, remember the side you cut depends on the pattern of carpet being used. If installing cut loop-pile, use a utility knife to make a path through a row of pile, then cut through the front of the carpet making sure not to cut through the actual pile. When cutting all other types of carpet, cut from the back side. Remember to avoid cutting through the actual carpet fiber.

about 30 seconds. Press the two edges of carpet down onto the tape. Remember, this is extremely hot so be cautious when pressing down. Make sure the fibers are not crushed down between the seam so it looks like a smooth, consistent piece of carpet. Place some weight on top of the seam while it sets. If there are any gaps still showing, you could take your knee kicker and gently push the seam closer together while the glue is still hot. Follow this same procedure for all the seams you have, including seaming the smaller pieces into the closets.

Once all the seams are completed, you are ready to stretch the carpet. There are a number of ways you need to stretch the carpet in order to ensure there are no ripples or bumps.

If the room is a space with four corners, A, B, C, and D, you can follow this method of stretching. Lift up and hook the corner of the carpet onto the tackless strip in corner A. Move back to corner A and stretch from A to B. Now move back to corner A and stretch carpet from A to corner C. Next stretch the carpet from A to D (Figure 1.13). If the carpet looks distorted or the pattern does not match, restretch the whole carpet again.

Use your power stretcher and the extensions, if necessary, to tighten and stretch the entire carpet. First go along the entire width, wall to wall with the stretcher until it is pulled into place onto the tackless strips. While the stretcher is still in place, use your knee kicker to finish attaching and securing the carpet to the strips.

Use this method until all the carpet is secured to the tackless strips. When finishing the closet or extra areas, use a perimeter wall to place the stretcher against, then place the other end into the closet. Preferably, the back wall of the closet should be secured first, then finish up with the side walls (Figure 1.14).

After everything is secured onto the tackless strips, go around and cut out any vent openings using your utility knife. You can apply

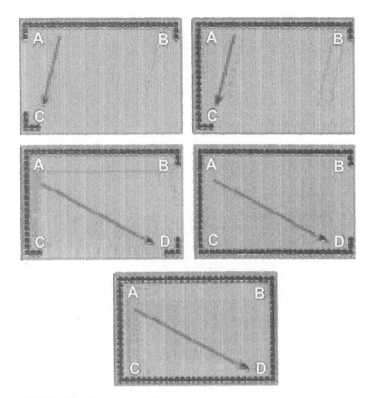

FIGURE 1.13 Carpet stretching chart.

double sided tape to tack down those loose edges or simply replace the vent covers over the carpet edges to hold the carpet down. Next, go around the perimeter of the room and cut off any excess carpet up to the edge with your carpet edge trimmer. Once that is completed, go back around and tuck in the trimmed edges with a stair tool or a putty knife.

INSTALLING CARPET ON STAIRWAYS

Stairway carpeting can be installed either by using one full strip of carpeting the same width as your stairway or by piecing it one step at a time. A single piece of carpet is best, so there are no seams showing which do not have to be tucked in at every step. For most

FIGURE 1.14 Carpet stretching.

installations, individual pieces that are the depth of the top of the stair and down the front of each riser are used. In both cases, measure the width of the stairway along with vertical rise and run of each step to determine how much carpet you will need. Take the carpet to a flat area and cut the appropriate size needed using your utility knife. Measure the length of each step and cut tackless strips for each stair. Cut and place them, one at the bottom of each riser and one at the back of each tread (Figure 1.15).

fastfacts

Before making the actual seam on the carpet being installed, make a test hot-melt seam using small scraps from the job. Allow the seam to cool, then cut a piece one inch wide by 12 inches long. Find a flat surface and stretch the seam over it by hand. If the carpet is going to peak, you will see it happen in your demonstration. This will show you whether you've done it correctly or not.

FIGURE 1.15 Placement of tackless strips.

Then, for each step, cut the padding and staple it to the total area being carpeted. Be sure not to cover the tackless strips. Make sure the padding is secure and flat before placing the carpet down.

Once the carpet is measured and the padding is installed, start at the bottom of the stairway and secure the carpet between the floor and tack-less strip. Using the knee kicker tool, take the carpet and stretch it up to the next tackless strip on the first tread. Take the carpet tool and tuck in the excess carpet between the tread and riser, kicking both sides until it is secured (Figure 1.16). Repeat this process for each stair.

Keep in mind that for stairways, you do not need to seam the carpet pieces together, just make sure that the carpet gets secured to each tackless strip and is tucked into each and every corner. When you reach the top of the stairway, there will either be another piece of loose carpeting or another type of flooring. If there is a loose piece of carpet, take the upper piece of carpet first and secure it

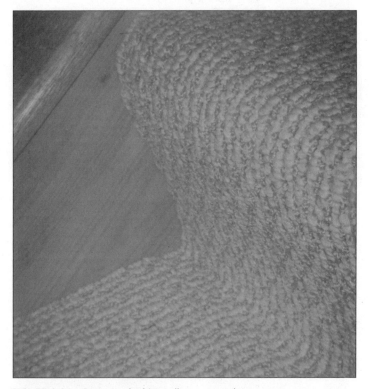

FIGURE 1.16 Carpet tucked into all corners and seams.

down, then stretch the lower piece so it meets the upper piece and secure that in place over the tackless strips.

If working on a stairway that is not enclosed on both sides, cut and wrap the carpet over the edges and secure it. When cutting the carpet, be sure to leave excess carpet for cutting and wrapping. Once the riser and treads are stretched and secured, go back to the edge and cut into it so that the carpet can go around any spindle or railing that exists. Once all the obstacles are cleared, fold the carpet down over the side edge first. Staple the carpet down and proceed to the other loose edges and secure them down with the staple gun (Figure 1.17). Now remove any excess carpet by cutting it with the utility knife. Repeat this same process for each edge.

Carpet installation can be a somewhat difficult job. Give your crew and yourself time and energy to get the project organized and installed properly so you and your customer can enjoy the benefits for a long time to come.

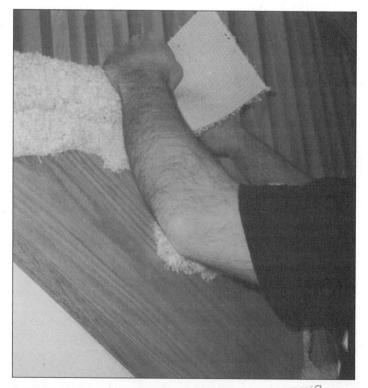

FIGURE 1.17 Wrapping carpet over stairway edges.

LEEDS COLLEGE OF BUILDING LIBRARY

LAMINATE FLOOR INSTALLATION

L aminate wood floors are quickly becoming the first choice in floor covering. Consumers are finding out that they can have the look of real hardwood for less cost and take less time for installation. These options not only put a new floor in one room, but make it feasible to place it in rooms the customer may not have been able to before. The new laminate floor products are far more resilient than the real hardwood and are easier to maintain. It can add beauty to any room in the house.

LAMINATE CHOICES AND CONSTRUCTION

As with any flooring product, thoroughly investigate the options available. In regards to laminate floors, there are two basic styles to choose from, either the planked boards or parquet square tiles. Both are available in a nice range of stains or unfinished, to be stained or painted a custom color (Figure 2.1).

Most hardwood flooring comes in strips, ranging in size when it comes to thickness and width. The thickness of the strips vary between $\frac{5}{16}$ inch to 1½ inch and the widths are between ¾ inch to 2¾ inch. Flooring can be purchased in single strips in different or the same lengths. Almost all brands of laminate flooring are now sold in strips which are connected at three pieces wide (Figure 2.2).

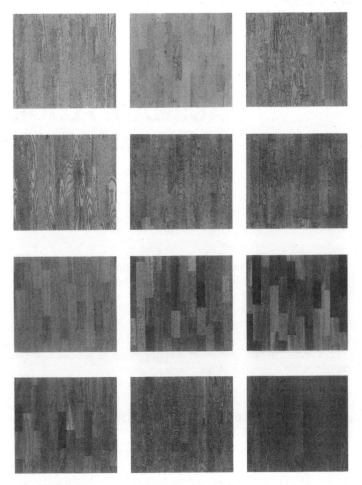

FIGURE 2.1 Stain options for laminate wood floors.

The three pieces are the same width and thickness to help speed the installation process along. Remember the more individual strips you use, and the smaller they are, the more time consuming and laborious the job will become. Laminate floor strips also come with each piece tongue and grooved, which allows for an easy, tight fit (Figure 2.3).

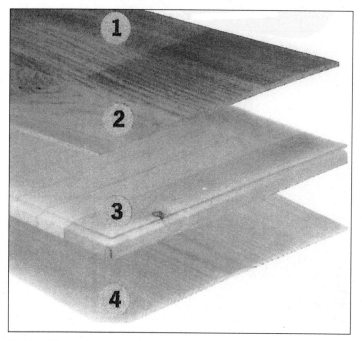

FIGURE 2.2 Construction of laminate floor planks.

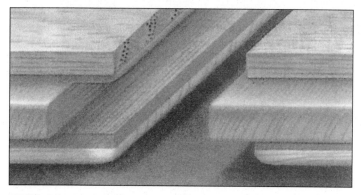

FIGURE 2.3 Tongue-and-groove edges.

SELECTING A FLOOR LOCATION

Once the appropriate floor is found, where it is installed becomes your next decision. Laminate floors can be installed almost anywhere. Unlike carpet and ceramic floors, laminate floors have few drawbacks. In kitchens where there is heavy traffic and constant stain possibilities, the ease of cleanup and the resistance to scratches from appliances and shoes make laminate floors a natural choice.

Although family rooms have traditionally incorporated carpet as the choice for flooring, there is a current trend toward laminate hardwood in the family room. A heavy traffic flow and lots of activity abound in this area, increasing the appeal of laminate hardwood floors that are easy to keep clean and add a touch of warmth to the living area.

Even the bedrooms are getting into the act of hardwood floors, which can add warmth and long-lasting beauty, rather than the alternative of worn-out and stained carpet. The simple ease of cleaning with an oil mop, rather than vacuuming and twice-yearly carpet cleaning, makes this flooring type a top option over carpet.

In a bathroom setting, whether it be a master bath or powder room or even a laundry room, laminate floors make sense. They are water resistant, hold up under heavy traffic, and still look great after a simple cleaning.

When selecting the location for the floor, be aware that since this type of flooring is made from a natural product. Whether it is a real piece of wood or a laminate plank made from real wood product, it can fade over time. So if you place furniture or rugs over a certain section of the floor and direct sunlight hits the area around it, you may see some distinct patterns on your floor (Figure 2.4). Periodic rearranging of furniture used in the room, or drapery treatments on the windows that block direct sunlight, can help slow down the process of fading.

In all of these examples, when it comes to cost, laminate flooring is still the better bargain. Even when dealing with the sub-flooring, plywood or cement, this is a less costly project and will last longer than carpet after being properly installed. These are just a few examples of options and reasons why you should consider laminate floors for the home.

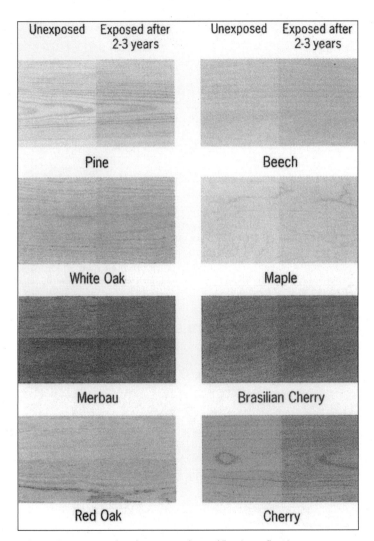

FIGURE 2.4 Exposed and unexposed wood/laminate flooring.

MEASURING FOR LAMINATE FLOORS

Once you've decided what type of floor you are using and where it will be placed, you are ready to measure. If you are using the planks, measure the room at a right angle to the direction you are installing the planks. If the room width is not an exact multiple of the width of the plank, plus allowing for the ¼ inch expansion distance, you will need to cut the last row along the long side of the plank making it narrower than the other rows. The last row must be at least 2 inches wide. If that is not possible, you must reduce the width of your first row (Figure 2.5).

If installing the square parquet flooring, use the chart in Figure 2.6. Find the length and width of your room on the top and side of the chart, following them to where they intersect. That number is the amount of boxes needed to cover the total area. Just as with the planks, allow an additional five percent for cuts and mistakes.

In both cases, strips or squares, when encountering an unusual shaped room, divide the room into several rectangular shapes and calculate each shape, then add the boxes together to come up with the total needed.

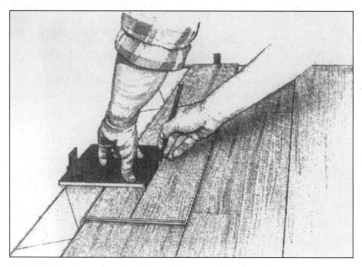

FIGURE 2.5   How to reduce last or first row to fit properly.

	2	3	4	5	6	7	8	9	10	11	12	13	14	15	16	17	18
2	1	1	1	1	1	1	2	2	2	2	2	2	2	3	3	3	3
3	1	1	1	2	2	2	2	2	3	3	3	3	3	4	4	4	4
4	1	1	2	2	2	2	3	3	3	3	4	4	4	5	5	5	5
5	1	2	2	2	3	3	3	4	4	4	5	5	5	6	6	6	7
6	1	2	2	3	3	3	4	4	5	5	5	6	6	7	7	7	8
7	1	2	2	3	3	4	4	5	5	6	6	7	7	8	8	8	9
8	2	2	3	3	4	4	5	5	6	6	7	7	8	9	9	10	10
9	2	2	3	4	4	5	5	6	7	7	8	8	9	10	10	11	11
10	2	3	3	4	5	5	6	7	7	8	9	9	10	11	11	12	13
11	2	3	3	4	5	6	6	7	8	9	9	10	11	12	12	13	14
12	2	3	4	5	5	6	7	8	9	9	10	11	12	13	13	14	15
13	2	3	4	5	6	7	7	8	9	10	11	12	13	14	14	15	16
14	2	3	4	5	6	7	8	9	10	11	12	13	14	15	16	16	17
15	3	4	5	6	7	8	9	10	11	12	13	14	15	16	17	18	19
16	3	4	5	6	7	8	9	10	11	12	13	14	16	17	18	19	20
17	3	4	5	6	7	8	10	11	12	13	14	15	16	18	19	20	21
18	3	4	5	7	8	9	10	11	13	14	15	16	17	19	20	21	22

FIGURE 2.6 Calculation chart for parquet squares.

33

Measuring for Laminate Floor Patterns

Standard laminate planks and parquet squares look great in their own right, but did you know you can cut the planks or even the square in any dimension you want and come out with a beautiful pattern for your floor?

There is a simple formula for figuring out a custom pattern if you are using planks. First, count the total number of planks in one complete pattern repeat, then multiply this number by the number of square footage in each plank. Now add the total square footage in one complete pattern repeat. Divide the square footage for each individual plank by the total square footage per pattern repeat. This will give you the percentage of the pattern that will be covered by each plank. Be sure to add five percent to the total square footage for cutting allowance. Divide the total square footage per plank by square feet per carton to determine the number of cartons required of each plank, then round up to the nearest full carton.

Standard size planks can be custom cut to your own dimensions or purchase product with laminate planks pre-cut to specific dimensions. Most pre-cut product will include four fixed lengths, making it somewhat easier to achieve their designs. Check out (Figure 2.7) for ideas on pattern design and layout.

LAMINATE FLOOR PREPARATION

Once the design layout has been determined and the correct amount of flooring has been purchased, lay the planks flat in their unopened boxes at room temperature for a minimum of 48 hours, acclimating them to room temperature. If the area is very dry or humid, allow up to an additional 48 hours for this process. Remove any transitions and moldings between and around the room where installation is taking place.

Before starting the job, inspect each piece for knots, discoloration, or other damage, rather than finding out in the middle of the project some pieces need to be replaced. Make sure each box includes the correct finish on all the planks or squares. What follows is a checklist of the proper tools needed to begin the installation of a floor.

Tools Needed for Laminate Floor Installation

- ✔ Measuring tape
- ✔ Chalk
- ✔ Straightedge
- ✔ Power drill and bits
- ✔ Hammer
- ✔ Nail set
- ✔ Power nailer
- ✔ Rubber headed mallet
- ✔ Pencil
- ✔ Clamps
- ✔ Handsaw or power saw
- ✔ Finish or flooring nails

Nail sizes should be 4d or 5d finishing nails if the flooring is up to ⅜ inch thick, 5d or 6d finishing nails on strips ½ inch thick, and 7d or 8d flooring nails on strips ¾ inch thick.

DETERMINING YOUR SUB-FLOOR

Depending on where installation of the laminate floor occurs, it should be either on a concrete or plywood sub-floor. For either type of sub-floor, make sure it is completely clean, dry, and flat. To make sure the floor is level, run a line of string across the room and place a level over it, or take a straight piece of long wood, six feet or longer, and lay that over the surface of the floor.

Concrete Sub-floors

If using a concrete sub-floor, I recommend using a moisture barrier between the concrete and the sub-floor. Even if a concrete floor is covered by vinyl, linoleum tile or sheet flooring, a moisture barrier should still be used. The thickness of the polyethylene film membrane should be between 0.2 mm and 0.15 mm thick. This should stop any water from soaking into the wood and damaging the new floor.

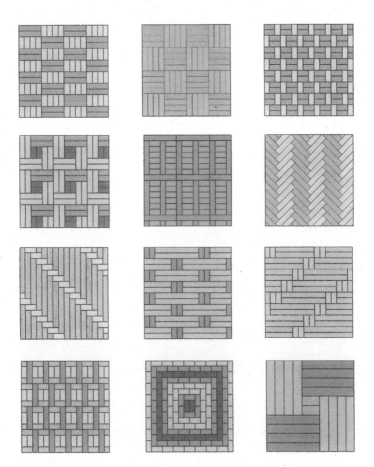

FIGURE 2.7 Patterns and design layout ideas for pre-cut laminate planks.

Before laying the moisture membrane, make sure that any low spots greater than ³⁄₁₆ inch are filled in with cement filler. A material called leveling compound, sometimes known as screed, can be used to level the entire floor out. For each 1mm of screed used, allow one day of drying time before installing the laminate flooring. Make sure there are no ridges or high spots in the floor. If there are high spots, they must be smoothed out with a grader or sander before installation of the sub-floor can occur.

fastfacts

Before going over a concrete sub-floor with a laminate floor, here is a quick way to perform a moisture test. Cut a couple of 2 foot squares of plastic and duct tape them to various areas of the concrete floor. Wait 72 hours and check for moisture. If there are beads of moisture on the underside of the plastic, there is a moisture problem. Even if the floor is dry at the time of installation, be sure to lay down a moisture barrier before installing underlay.

Wooden Sub-floors

As with any sub-floor, make sure the wooden sub-floor is flat, dry, and clean. While sweeping it clean, check for nails, staples or tacks from previous flooring installations. Don't forget to check for any large holes like knots or gaps in the wood. These must be filled in with wood filler if they are larger than $\frac{3}{16}$ inch. If there are any loose boards in the sub-floor, use screws rather than nails to secure them down. You may want to use a detector to make sure there are no pipes or wires underneath the floor when driving screws or nails into it. Use a plane to smooth any high spots in the floor.

TYPES OF UNDERLAY FOR LAMINATE FLOORS

When the sub-floor is leveled off and cleaned, an underlay must be installed. In general, there are three different types of underlay that can be used. Basic laminate flooring underlay can be used with all types of laminate floors. A wood fiber underlay should be used if the job requires extra acoustic or thermal properties. This would work best in a basement or family room with a stereo or home theater system. Last, a silver insulating underlay should be used to prevent moisture and rot. Remember, laminate hardwood floors cannot be directly adhered to a sub-floor. Be sure to choose the appropriate underlay to go between the sub-floor and the flooring.

fastfacts

Before the underlay or flooring goes down, make sure what you are nailing or screwing into first. To find out if there are PVC pipes or radiant heat underneath, spray mist some water on the floor and turn up the heat. Look at where the floor dries fast and first. This could show a pattern of the pipes for heating underneath and where not to nail when installing flooring material.

CUTOUT AND TRANSITIONS

When installing the new floor, no doubt obstacles will come up, such as pipes, outlets, etc. When cutting the laminate floors, use a power circular saw with carbide toothed blade, cutting from the non-decorative underside of the board to avoid chipping. When cutting from the tops of the boards, use a hand saw. Always cut the planks carefully in order to avoid chipping or splitting.

One common cut-out area is under a door frame. You want the flooring to go under the frame to give it a professional look and cover any uneven edges. Simply take a piece of the floor and lay it upside down next to the door frame. Make a mark on the frame and saw off the bottom of the door frame. Slide the flooring plank or square underneath the frame, making sure to leave enough room for expansion once the floor is finished (Figure 2.8).

When cutting out openings for pipes or vents, approach them from either the short end or long side of the plank. If using the short side, measure the diameter of the pipe or opening you need to go around. To leave room for expansion, make the opening an additional ½ inch larger than needed. Take a drill and place it at the center of the hole and go all the way through the plank or square. To complete the fit, cut the plank through the center of the hole (Figure 2.9).

When cutting on the long side of the plank, drill openings the same way as the short side, except when completing the fit, and cut the plank at a 45 degree angle instead of through the center (Figure 2.10).

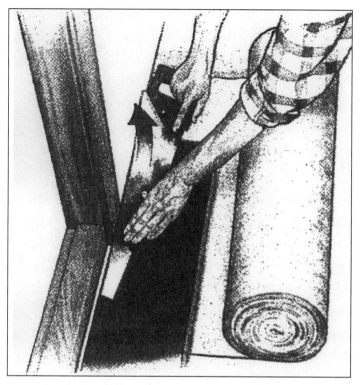

FIGURE 2.8 Fitting the underlay under a door frame.

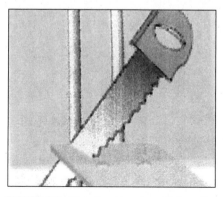

FIGURE 2.9 How to cut an opening on the short side of the plank.

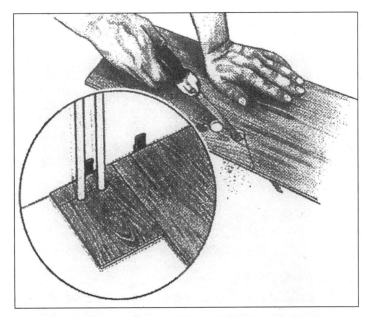

FIGURE 2.10 How to cut an opening on the long side of the plank.

INSTALLATION OF LAMINATE FLOOR PLANKS

This section is about installation of laminate floor planks with either nail or glue methods. The next section will discuss how to lay out and install laminate floor squares.

The location of the floor and the type of sub-floor used will determine the method of installation. The nail or staple method can be used if installing the floor over ½ inch or thicker plywood or over an existing wood floor. The glue down method can be used when installing the floor over plywood, concrete, particle board, existing plank sub-floors, tile or existing vinyl sub-floors.

Installation of Laminate Floors using the Glue Method

Before starting the actual installation of the laminate flooring, decide which direction the floor will go. It is usually standard practice to run the planks lengthwise. If there are windows in the room, run the

fastfacts

*When encountering any type of cut-out, a cardboard template of
the size and shape can be made. Use this as a pattern, placing it
over the piece to be cut. This will give the most accurate shape of
the obstacle to be cut around.*

planks toward them, parallel to the incoming light source. If laying
over existing floorboards, run the planks at a 90 degree angle to the
existing floorboards. This helps to reinforce the joints. Start in the
left corner of the room and work from left to right, dry running the
first three rows of planks. Make sure the side of the plank with the
groove is facing the wall. Working from left to right, leave a space
on the outside of the wall for expansion.

A ¼ inch expansion space is standard unless the room is 66 feet
or greater in length or width, then an expansion of ½ inch is required
around the entire perimeter of the room and any fixed objects such
as pipes or supports. Use spacers available from your supplier to
maintain consistency throughout the installation. After installation of
the planks is finished, cover any expansion spaces with the molding
and quarter round.

The first row should be started with a full size length plank, the
second row should be two-thirds the length of the first plank, while
the third plank should be one-third the length of the first plank. This
creates a natural, staggered pattern throughout the room. After dry
fitting the planks, you will be able to see if the wall is straight (Figure 2.11). If it is not, take a pencil and make a line along the first
row of planks, making sure to use spacers and maintaining a consistent space away from the wall. Once the planks are placed straight
with the wall cut in a staggered pattern for every three, they are
ready to be glued down.

Use a saber or circular saw to cut the planks then number them
on the back according to the sequence they are installed. You can
now glue and assemble the planks according the number sequence
on the back of each piece. Place glue in entire groove and on both
the long and short sides of the plank (Figure 2.12). You do not need
to apply glue where the plank faces the wall or around any fixed
objects in the room. Now place a wood block against the side of the
last plank glued and tap it snugly in place with a rubber mallet. A little

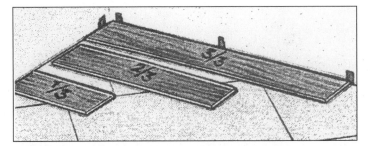

FIGURE 2.11 Dry fitting the first three planks.

glue will probably come out of all the edges around the gaps. That is natural and it shows that enough glue has been placed in the grooves. Wipe off any excess glue right away with a damp cloth. When you are finished, there should be no gap between the planks. If there are gaps, keep tapping with the rubber mallet and block of wood until all gaps are eliminated.

After installing the first three rows, give the glue about an hour to dry and set before installing the rest of the floor. After the floor has had time to set, continue installing the floor, working from left to right, plank by plank, row by row. Be sure to keep the pattern consistent as you go, staggering the planks like the first three. Make certain there is a distance of at least 8 inches or greater between joints from one row to the next (Figure 2.13).

When you reach the last row, you may need to cut the planks lengthwise. No room is ever perfectly square and when you reach that last plank, it must be spaced the same distance away from the wall as the first plank.

Take a full length of flooring and place it over the last row of planks you installed. Line up the long side, making sure the joint is a minimum of 8 inches apart. Trace the contour of the wall on top of the last strip to place along the wall. Cut along the line accordingly, then glue and fit the last plank in place (Figure 2.14).

Leave all the spacers in place along the wall until the glue is completely dry. Let the floor dry for at least 12 hours before cleaning or using. If the floor gets wet before it is completely dry, water may get into the cracks and create warping. If the floor is used too soon, pressure from walking on top may create gaps between the planks or an uneven floor. Check for any excess glue or marks and remove right away with a damp cloth. If that doesn't take it off, try cleaning with a solution of a ½ cup of ammonia in one gallon of water.

FIGURE 2.12 Apply glue to the planks.

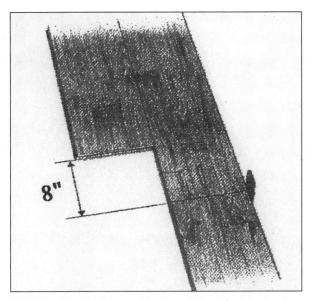

FIGURE 2.13 Maintain an 8 inch distance of joints from row to row.

FIGURE 2.14 How to measure and cut the last plank.

After the floor is set, remove all the spacers around the perimeter of the room and install the trim or quarter round that should cover the expansion space around the perimeter. Install the transitions and apply any sealants necessary. Sealants should be applied between the floor and dishwashers, sinks, kitchens in front of dishwashers, sinks, laundry rooms, and exterior sliding glass doorways.

Installation of Wood Floor Planks Using the Nail Method

The nail down method for installation of planks is best suited for a floor going over another wood floor and has at least a ½ inch plywood sub-floor between them. Begin by using a chalk line and a straightedge, marking out a ½ inch space all the way around the perimeter of the entire floor. Lay out the first row of strips along this line, making sure the grooved edges are facing the wall (Figure 2.15). Using the same rules of layout discussed in the previous section, dry lay the strips lengthwise, working from left to right. Place the planks along the chalk guideline and face nail, straight down, every 8 inches.

Nail close to the groove so that all the nail holes will be covered when the trim is replaced. Be sure to pre-drill the holes so as not to split the wood. Use 4d or 5d finishing nails on wood strips up to ⅜ inch thick, 5d or 6d finishing nails on ½ inch thick strips and 7d or 8d flooring nails on strips ¾ inch thick.

After placing and nailing down the first row of wood strips, you should see the tongue edge of the wood plank exposed. Drill holes at a 45 degree angle into the tongue every 8 inches. This will prevent

FIGURE 2.15 Alignment and nailing of starter strip.

the wood from splitting when nailing into it. Place a nail in each hole and nail it in halfway, go back and use a nail set to hammer them in the rest of the way, until the nail head is slightly below the surface of the floor.

As with the laminate floor, set a course of correctly cut and sized pieces, marking them on the back according to where they are to be placed. Follow the numbered pieces and install the planks piece by piece. As you move away from the perimeter walls, begin using a power nailer to nail the strips down (Figure 2.16). This makes the job much easier and quicker.

Revert back to the hand nailed method as you get closer to the end. Follow the cutting method described for finishing the laminate floor. Remove any spacers located around the perimeter of the wall. Replace any molding or transitions, installing new product if needed (Figure 2.17).

INSTALLATION OF LAMINATE TILES

Begin installation of tiles by opening at least three boxes of the tiles and mixing them together. This blends different shades and styles, giving the floor a more natural appearance. When it comes to balancing the design and pattern, tiles are a little more complex. The layout is basically done the same way as planks, starting at the left side of the longest wall. The tiles can be aligned with either the four corners meeting or in a staggered style. Placement should be balanced whether it is in a hallway or a full size room. You don't want to end up with a full tile on one side of a hall and a half width tile on the left side.

Lay out a row of tiles running the length of the room, then measure the distance between the last full tile and the wall. To create balance, add that distance plus the width of the tile. Divide that sum by two and that number will be the size of the tiles needed to balance that row (Figure 2.18).

Use that same method for determining the width of the row of tiles needed. Once the distance away from the perimeter of all the walls has been determined, mark that parallel line along the longest wall with a chalk line (Figure 2.19).

Dry lay out two rows of tiles, finding out where they will meet at length and width. That is the starting point. Place a tile there, with one groove side to the left and the other groove side against the chalk line (Figure 2.20).

FIGURE 2.16 Power nail method.

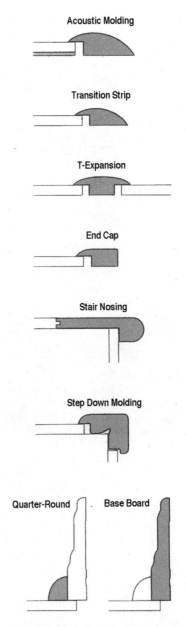

FIGURE 2.17 Transition and baseboard options.

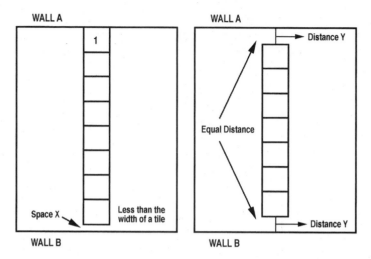

FIGURE 2.18   Layout of tile row creating equal length balance.

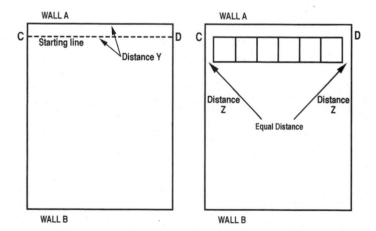

FIGURE 2.19   Layout of tile for equal width and chalk line placement.

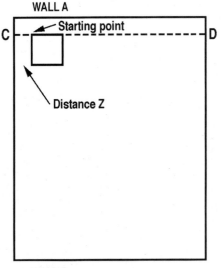

WALL B

FIGURE 2.20 Finding the starting point.

Lay out the first row of full size tiles, then go back and cut the half size tiles. Place a tile on top of the full tile, then place a second tile on top of that one, always making sure that the grooves and edges match the direction of the tiles below it. Push the tiles against the wall with a spacer measuring ¾ inch between the wall and the two tiles. Make a mark on the top tile; this will give the exact size tile measurement needed to fit that space. Use a carbide tip saw blade to cut tiles, being careful not to chip the tile. Repeat this process along the walls where tiles smaller than full size are needed. Keep an expansion space of at least ¾ inch around the entire perimeter of the room. Fit and make all of your cuts ahead of time for a few entire rows. Since the tiles have to be glued as you go, they'll already be precut and ready to glue. Work row by row and remember, when gluing, to fill the entire groove with glue, tapping the tiles gently with the wood block and rubber mallet to assure a secure fit. Wipe off any excess glue around the edges with a damp cloth.

After installing the first two rows, allow one hour before you install the rest of the floor. Repeat the same steps until all rows are installed. Cut out any openings necessary the same way as described in the laminate installation section. Once the glue has been allowed

fastfacts

Parquet tiles are usually composed of odd shapes and cutting them requires special care. Clamp the wood to a flat extended surface and cut with a saber saw. To prevent chipping, use masking tape on the edges to be cut. Redraw the line over masking tape and make the cut. The tape supports the edge and minimizes splintering and chipping.

to dry and the floor has set at least 12 hours, remove the spacers around the perimeter of the room. Replace the molding and transitions where needed. Check for any excess glue around the edges of each tile and wipe away with a damp cloth. If the glue is already set, mix a solution of ½ cup ammonia with one gallon of water and mop over the area until the glue is gone.

During the first six to eight weeks after installation some swelling may occur along the seams of the planks or tiles. This is normal and actually shows that enough glue has been used in your installation. The swelling is caused by the absorption of glue into the core material of the floor. This swelling will disappear after the glue fully cures.

As with any installation, careful planning and the right tools are essential. When it comes to laminate and wood flooring, it is especially important to start perfectly square and number the pieces as you go. This can make the project go quickly and smoothly.

3

CORK FLOOR INSTALLATION

C ork is a simple, inexpensive, and natural alternative to the more well known flooring options. This fairly new flooring source is available in rolled sheets, 1 x 3 foot planks, or square tiles. Cork is also available as an underlay for other flooring materials such as laminate, hardwood, ceramic and stone floors.

PROPERTIES OF CORK FLOORING

Cork flooring is manufactured from the bark of the cork oak tree. This type of flooring offers great acoustical properties, dampening the sound within the room. Since cork is a natural material, it does not omit any gases or micro fibers, like other by-products from various flooring sources. Cork is also a great insulator, warm and soft to walk on. Since cork is a somewhat porous material, it does not entrap dirt or fungus. It also does not rot and is easy to clean, making it a very hygienic product to use.

VARIETY OF CORK PRODUCTS

Cork flooring is available in one by three foot planks, consisting of three layers. The surface layer is 100 percent cork with a UV-cured acrylic finish on top, making it easy to clean and longer lasting. The middle layer is a pre-cut tongue and groove construction, making it an interlocking system not unlike most laminate plank floor products. The base layer is made up of cork underlay, providing better acoustical properties and insulation (Figure 3.1).

Cork flooring is also available in square tiles both for the floor and the wall. You can order any type of cork flooring in a variety of colors and patterns (Figure 3.2). It is usually available finished with the standard or custom finishes, but you can order it unfinished if desired.

DECIDING ON A LOCATION FOR CORK FLOORS

Placement of this type of flooring can be in virtually any room of the house or office. If you are seeking to lower the noise level in a room, the family room is a great area to install cork. Even in heavy traffic areas, like foyers, kitchens, or laundry rooms, cork is a reliable product to use. It does not rot or collect fungus when it gets wet or dirty and it is easy to maintain. Remember, cork is similar to a wood floor in that it will fade over time. Be sure to install it in an area protected from direct sunlight.

CORK PRODUCT AVAILABILITY

Cork flooring is available in four types of product, ready for installation. The first type of cork is unfinished, stained and ready to urethane (Figure 3.3). This cork comes in tile form with a stained

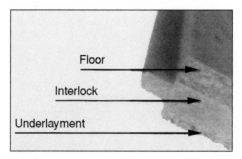

FIGURE 3.1 Layers of cork floor.

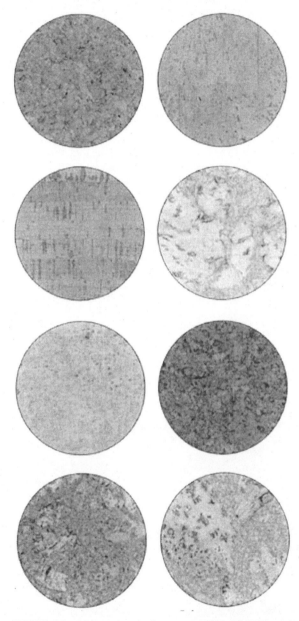

FIGURE 3.2 Many colors and patterns are available for cork flooring.

surface and requires three to four coats of low satin urethane to seal it, making it ready for use. The cork surface is semi smooth with a visible natural texture, either with a square or beveled edge. Tiles are pre-glued on the back with water-based, contact adhesive. For tiles made of cork, you must apply adhesive to the sub-floor before laying the floor.

Another type of cork product available is unfinished, natural and ready to urethane (Figure 3.4). This tile comes with a natural unfinished surface, and, like the other tiles, must have three to four coats of urethane applied to it for sealing purposes before it is ready for foot traffic. It also comes with pre-glued backs and must have adhesive applied to the sub-floor first before installation of the tiles can occur.

A third type of cork is pre-finished and urethane covered (Figure 3.5). This cork is pre-finished with several coats of low gloss urethane, giving it a smooth finish. Pre-finished tiles are only available in the square edge style. These tiles are also pre-glued on the back and must have adhesive applied to the sub-floor before installing the tiles.

The fourth type of cork product is pre-finished vinyl-covered flooring (Figure 3.6). The vinyl is fused to the surface and sealed with acrylic lacquer to prevent moisture seepage. The surface is smooth with a low gloss finish. This product is not pre-glued on the back, but the sub-floor still requires adhesive glue before installation of tiles. An optional cork dressing may be applied to this vinyl surface, making it easier to clean spills and adding an extra layer of protection.

FIGURE 3.3 Unfinished stained cork.

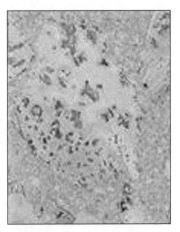

FIGURE 3.4 Unfinished natural cork.

FIGURE 3.5 Pre-finished urethane covered tiles.

PREPARATION OF THE SUB-FLOOR

Installation of the cork flooring can be made over either concrete or plywood sub-floors. Most cork product can be applied directly to an existing sub-floor. An option for the floor is to use an underlay of cork. This product adds an additional 50 decibels of sound reduction, making a room a little more quiet and insulated.

Concrete Sub-floor

Cork floor can be installed over any concrete that is above grade, grade level or below grade. The concrete should be smooth, clean

FIGURE 3.6 Pre-finished vinyl covered tiles.

and level to ⅛ inch thickness within every 10 square feet of surface. Fill in any low areas with screed or leveling compound and grade down any high spots. After filling and leveling compounds have dried, cover the concrete with 6 millimeter polyethylene sheets. Overlap the sheets at the seams a minimum of 8 inches and tape them together, assuring a tight moisture barrier.

Plywood Sub-floor

Cork flooring may be installed over plywood floors much the same way it is over concrete. Make sure the floor is smooth, clean and level. Sand down any high spots and apply wood fill to any holes or low areas. Allow the filler to dry and check for moisture content.

fastfacts

If moisture content is above 14 percent, you should install poly-ethylene sheets over plywood before laying cork floor. If the room is over a crawl space or slab, it is best to install the moisture barrier, avoiding future damage from moisture.

INSTALLATION OF CORK FLOOR

In this section, let's discuss how to install both cork planks and tiles. Here is a checklist of tools needed for this project:

Tools for Installing a Cork Floor

✔ Table saw

✔ Tape measure

✔ Spacers

✔ 6 mil film of polyethylene

✔ Polyurethane cleaner

✔ Pencil

✔ Cork floor squares or planks

✔ Chalk line

✔ Adhesive glue

✔ Knocking block of wood

✔ Optional—Polyurethane (low gloss) sealant

✔ Optional—Sub-floor adhesive and applicator

✔ Optional—100-pound floor roller (local rental or hardware center)

PREPARATION OF CORK BEFORE INSTALLATION

As with most natural floor products, cork flooring must become acclimated before installation. Remove the cork from its wrapping and place in the room where installation is planned for a minimum of 72 hours. Maintain the heat, humidity or air conditioning levels consistent with the normal room environment during this time.

Remember, cork is a natural product, and just like wood, cork tiles and planks will vary in shade and pattern. Cork is also prone to the same type of expansion rates as wood. Minor expansion will occur during periods of high humidity and minor contraction during low humidity. This happens during seasonal changes as a room gets dry in the winter or has high humidity during the summer.

Cork Plank Installation

Cork planks do not require adhesion to the sub-floor. These planks are held down by their own weight and are held together by tongue-and-groove construction only.

To begin, as with any flooring installation, remove any existing moldings and transitions from around the room. Now, find the longest wall that will be parallel with the planks. Measure out ½ inch away from the wall all the way around the perimeter and make a chalk line. This allows an expansion area around the floor and sets your line for the first row of planks. Mix up the planks from at least three of the boxes to achieve a nice blend of color and pattern throughout the project. Now start at the left end of the room and work your way to the right, groove sides facing the wall. Place your spacers at ½ inch from the wall, two per plank, while keeping the plank lined up on the chalk line. The spacers will help keep the first row from shifting when adding the second row of planks.

You may want to dry set the planks to figure the layout first. The planks should be placed in a staggered pattern, just like any planked floor product, making sure that no joints are aligned with one another. Measure and cut accordingly, then mark them on the back in the proper order they will be used.

Once the layout is established, the pieces are cut and numbered for the first few rows, ready to install. Take the first numbered piece and apply glue to inside of the groove and any short ends where another plank will touch it.

When installing the second row, do not apply glue to the tongue side of the plank. Only apply glue to the groove side, then take your knocking block and gently tap the planks together. Do not use a hammer to tap the planks together, always use the knock block when tapping the pieces together. Once the planks are secured,

fastfacts

The best way to cut the cork is to use a fine tooth blade with a table saw. Place the cork plank decorative side down when cutting. This technique will help keep it from chipping or splitting on the finished side.

move on to the next piece, row by row, plank by plank, until you reach the end of the wall. When you reach the last row, you will most likely have to cut the planks lengthwise.

Follow the chalk line and cut accordingly, allowing for the ½ inch expansion space. Let the glue set for at least 24 hours then remove the spacers from the perimeter of the room. If any glue is found on the surface of the floor, it can removed with a rag dampened in mineral spirits.

Cork Tile Installation

Like the cork planks, the cork tiles must be acclimated to the room they will be installed in. Give them 72 hours to reach the same levels of temperature and humidity.

Cork tiles require a sub-floor primer to be applied before installation. Regardless of what type sub-floor being used, a natural and water-based sub-floor primer must be applied before installing cork tiles. Pour the primer into a paint tray, take a short-nap roller and apply a thin coat of primer onto entire surface. The primer should dry in about 45 minutes under normal temperature and humidity. Be sure the primer coat is completely dry before applying adhesive on top of it.

Apply the adhesive coat the same way as the primer, pouring some into a paint tray and rolling it on with a short-nap roller. Apply adhesive to an area of 50 square feet at a time. A consistent, glossy coat indicates a sufficient amount of adhesive has been applied. This coat will dry in about 20 to 30 minutes and then the tiles must be applied within one hour of adhesive application. If the sub-floor is very porous, a second coat of adhesive may be necessary.

When laying out the pattern for the cork tiles, it is best installed using the staggered joint method. Just like the cork planks, be sure to mix at least three boxes for the best mix of color and pattern. Allow for a ¼ inch expansion, placing spacers around the room for consistency.

Mark a chalk line ¼ inch away from perimeter walls and lay the tiles row by row, piece by piece. The tiles are already pre-glued and will adhere to the sub-floor quickly, so once the first tile is placed down, it will set the standard for the rest of the floor. To complete laying the floor, place the tiles next to one another and gently tap them together for a secure seam. It is not necessary to press tiles firmly together since there should be some room for natural expansion between each one.

fastfacts

Keep in mind that cork flooring is a very porous material. If the cork tile or plank is not sealed, it will absorb primer, water or anything liquid. Be careful not to set anything on the floor that is wet until you have applied all of the sealants. Let the primers dry on the sub-floor before laying the cork on top, otherwise the cork will absorb the primer from below and warp the cork.

After the entire floor has been laid, it must be rolled in several directions, several times, with a 100-pound floor roller. After setting overnight, the floor should be rolled again, several times in all directions, before the application of any sealer. In commercial applications, four coats of sealer are recommended. Residential cork floors should have a minimum of three coats of sealer applied.

After the final rolling is finished, let the floor set for at least 24 hours before using. Reinstall any molding and transitions where needed. If any excess glue is found on the surface of the floor, simply wipe off with a rag that has been dampened with mineral spirits.

In the case of either cork planks or tiles, it is recommended that an additional water-based polyurethane sealant be applied. Even though a factory coat of sealant is standard, unless special ordered as unfinished, an extra coat will only enhance the appearance and give unprotected joints protection from seepage, preventing darkness from dirt and water.

Cork flooring is a fairly simple and cost-effective option for any room of the house. Cork is quick and easy, once you are aware of the techniques for installation.

4

HARDWOOD FLOOR
INSTALLATION

Hardwood floors have quickly gained popularity with home-
owners over the past decade. Economically, hardwood flooring makes great sense. While carpet, tile and other flooring
products need to be replaced every five years or so, hardwood floors
last a lifetime. The initial cost quickly pays for itself with hardwood.
Wood can also be cleaned more thoroughly than carpet or tile, which
can trap dust and dirt. Also, hardwood floors can add strength to a
home and provide better insulation than other types of flooring.

A recent national survey of real estate agents states that 58 percent of homes with hardwood floors bring a higher price. A typical
installation of 500 square feet of hardwood flooring can add 5 percent to the value of a home. Installation of hardwood flooring is more
complex and less forgiving than a typical laminate floor, but, with the
right tools and experience, it can be done quickly and precisely.

TYPES OF HARDWOOD FLOORS

There are four types of hardwood floors:

- *Parquet*
- *Unfinished solid hardwood*
- *Pre-finished solid hardwood*
- *Engineered pre-finished hardwood*

Parquet

A parquet floor is made up of strips of wood made into a square (Figure 4.1). This type of hardwood flooring is the least expensive and is glued down to a sub-floor. Parquet floors are harder to refinish than other solid wood floors and the life span is relatively shorter.

Unfinished Solid Hardwood

An unfinished solid hardwood floor comes in non-varnished, rough strips (Figure 4.2) These strips have to be nailed down to a wooden sub-floor. A vast array of grades, wood species, and

FIGURE 4.1 Parquet floor.

FIGURE 4.2 Unfinished solid hardwood.

widths are available. This type of flooring is economical, however, staining and varnishing the floor can be costly, time-consuming, and complicated. Finishing this type of floor must be done at the installation site. When finishing the surface of this type of floor, it can take up to an additional three to five days to complete. The dust and fumes can also lead to vacating of the premises until all work is done.

Pre-finished Solid Hardwood

The pre-finished solid hardwood floor comes in strips which are already sanded, stained and covered with several coats of polyurethane (Figure 4.3). This type of floor is also available in several stains, species and strip widths. This is perhaps the easiest and quickest type to install.

FIGURE 4.3 Pre-finished solid hardwood.

Engineered Pre-finished Hardwood

Engineered hardwood floors were developed to be installed over concrete sub-floors (Figure 4.4). The strip of flooring is a piece of hardwood placed over a layer of plywood. The construction of the plank creates a piece which is the same thickness as a regular strip of solid hardwood, but is more stable to endure higher humidity fluctuations. These strips can also be glued or stapled down over a plywood sub-floor. The highest quality engineered floors are as good as, or even better than, regular solid hardwood and can be installed in any room of a home or commercial building.

SPECIES OF HARDWOOD

There are dozens of wood species to choose from, depending on the color and grain variations desired. The choice of species should be made strictly according to the color and pattern needed for an installation. The hardness of a species should not be a criterion in selecting a residential floor. The relative hardness of a species is determined by an industry test known as the Janka test. In the accompanying chart (Figure 4.5), the higher the number, the harder a species of wood is.

FIGURE 4.4 Engineered pre-finished hardwoods.

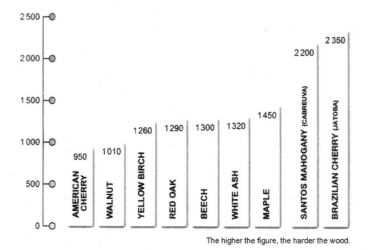

The higher the figure, the harder the wood.

FIGURE 4.5 Janka test chart.

Here are some suggestions for species selections. Maple is a good choice if you're looking for a subtle, natural grain appearance. This will produce a very pale, clear, and uniform look for the installation. If oak or white ash is chosen, you will see a very distinct color variation in the natural grain. Walnut and some of the exotic woods offer a darker appearance than the aforementioned species. A chart of hard and soft wood species typically used for floors is shown in Figure 4.6.

WOOD GRADES

The grade of wood is a technical term used to categorize wood according to its natural appearance. "Select and Better" grade has a more uniform color and less variations. If you see "Rustic" or "traditional" grade, these have a more pronounced variation in the grain.

WOOD COLORS

Pre-finished hardwood floors are available in a variety of shades and colors. Check with many manufacturers to see who offers the best color for your installation. Be sure to check for uniformity of color and that the stain is covering the entire surface, including the micro-v joints.

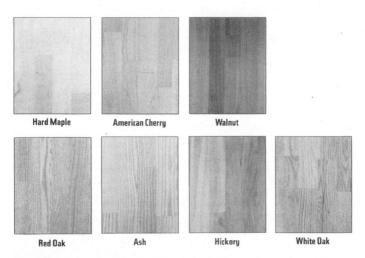

| Hard Maple | American Cherry | Walnut | |
| Red Oak | Ash | Hickory | White Oak |

FIGURE 4.6   Species of soft and hard woods.

fastfacts

When looking for flooring product, look at two samples of the same type of wood in two different grades. A noticeable difference should be evident. When the boxes of wood arrive for an installation, be sure to compare several strips within each box. Sometimes the manufacturer will give a good deal on the wood but will mix grades within the boxes. In the end, you will spend more because you will end up buying more product to make up for the defects.

Some exotic woods require a maturing period before the true color is revealed. If you inspect a box and notice a contrast in colors, the true color should become natural within a few months.

WOOD STRIP WIDTHS

The width of a wood strip can be between 2 inches through 3¼ inches (Figure 4.7). The width of the strip chosen will greatly determine the appearance of the floor. Also, the narrower the strip, the longer and more tedious the installation will be. If a narrow width is used, the pattern or grain of wood will also be less noticeable. A wide width will tend to highlight any graining or markings on the strip. Sometimes a combination of two widths can be used, but only if installed by a very experienced installer.

MICRO-V JOINTS

A component consistent with any well produced wood strip, either all wood or pre-engineered, is a micro-v joint. All strips must fit tightly and flush with each other, therefore, the fit must be concise. A micro-v joint is formed when two strips are put together. Once together a flat, tight fit should occur and the same thickness should be felt across the surface of all strips put together. The micro-v joint should not be too deep as to attract dirt, or making it hard to refinish if necessary.

FIGURE 4.7 Widths of wood strips.

The flushness of the floor will also keep wear and tear minimal on the edges and make it easier to move furniture around. Be sure to check the edges of strips for same thickness. A good test is to simply put a few strips together before installation and run your hand over them, checking for smoothness.

MEASURING THE FLOOR

Once the grade and width of strips are decided upon, the floor must be measured. To figure out a room's square footage, multiply the length by the width. If installing a floor in room with odd shaped areas, divide those areas into sections and calculate the area of each. Add all of the dimensions together to come up with the total square footage needed. Take the total number of inches and divide by 144 to find the square footage. Be sure to add 10 percent for waste and

fastfacts

If a wood floor is being installed in a room where there is a great fluctuation in humidity, narrow strips are the preferred choice. When the floor is affected by a change in temperature, there will be a larger number of joints to spread the movement over.

errors. Figure 4.8 shows a simple chart on how to calculate room area. Wood strips are sold in bundles which cover a specific amount of square footage. Wood parquet tiles have the amount of area specified on the side of each box.

TOOLS NEEDED FOR HARDWOOD FLOOR INSTALLATION

Here is a checklist of tools needed for installation of all types of hardwood floors and pre-engineered wood floors. This list includes tools for measuring, cutting and installation.

Tool Checklist

✔ Circular saw—This will work for removing old flooring and cutting new flooring

✔ Saber saw—This tool helps cut the edges off a pre-engineered board when reaching the end of a floor row

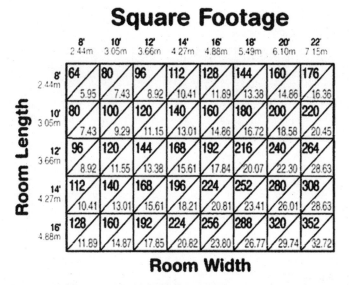

FIGURE 4.8   Room calculation chart.

✔ Electric drill—Drills pilot holes for screws into hardwood

✔ Drill bits—Needed to make holes for wood plugs or masonry bit for concrete sub-floors

✔ Power miter saw—Helps to cut boards at an angle when needed

✔ Jamb saw—Makes the cut under door jambs and trim.

✔ Level

✔ Chalk line

✔ Framing square

✔ Tape measure

✔ Butt chisel—Use this tool to pry off old floor and baseboard

✔ Utility knife

✔ Power nailer

✔ Hand plane

✔ Caulking gun

✔ Rubber mallet

✔ Claw hammer

✔ Nail set

✔ Regular straight edge trowel—Use for leveling compound

✔ V-notched trowel for adhesive

✔ Tapping block -wood

✔ Putty knife—Spreads adhesive onto sub-floor before it is v-notched

Not all of these items will be used for every installation. It all depends on the sub-floor being used and if the installation involves adhesive, nails or glue. Make sure these tools are nearby in case they are needed so a delay in the project does not occur.

SUB-FLOORS

There are three types of sub-floor used beneath a hardwood floor:

• *Plywood underlay*

• *Plastic foam*

• *Vapor barrier*

Plywood Underlay

When a floor is being installed over an existing wood sub-floor or a floor placed over a wood frame or concrete slab, a new ⅜ inch plywood sub-floor should be installed.

Glue down parquet floors can be adhered directly to a concrete floor; all other types of hardwood floors need a plywood sub-floor. A concrete slab can be easily raised by constructing a 2 x 2 inch wood grid. Take a power drill and make holes for the anchors which will attach the wood grid to the cement floor. Drill number 7 or number 9 screws into the wood through to the anchors. Once the grid is secured, place glue on top of the wood grid and lay the plywood on top. A thin layer of foam insulation can be placed between the 2 x 2 wood grids, depending on the climate.

Once the plywood has been laid, 1¾ inch shank nails should be used to nail it down to the wood grid. Lay the plywood using staggered joints, so none of the corners meet. Nail around the entire perimeter every 5 to 6 inches and on the 2 x 2 inch wood grid. Make sure to nail set all the nails below the surface of the plywood and fill in with wood fill, assuring a smooth surface.

If the project has an existing resilient sheet or tile floor, a layer of ⅜ inch plywood can be laid right over the top of it. Cut the plywood and lay it over the existing floor, after making sure it is level, smooth and clean. Any large gaps or holes should be filled in and left to dry before the plywood is laid. A gap of ⅛ inch between each piece of plywood is sufficient for expansion and contraction. Just make sure the corners of the plywood do not meet and to secure it down every 6 inches around the perimeter and seams. 2¼ inch nails or screws can be used to secure the plywood down. Once the nails are in, make sure they are set below the surface of the plywood and fill them in with wood fill.

Plastic Foam

Plastic foam sub-floors are primarily used for pre-engineered hardwood floors. No adhesive is applied to the floor, so the strips will float on top of the foam. This type of underlay is also a good sound reducer and adds to the insulation value of the floor.

Vapor Barrier

Before installing any sub-floor, run a test for possible moisture problems. The main enemies of a hardwood floor are moisture and humidity. Moisture changes can cause cracks, movement of floor, cupping and buckling. If you are placing hardwood floors below grade level, it is possible that you could experience moisture problems. Moisture can also enter a home through a laundry room or entryway. If the original floor had moisture problems, it either still exists or needs to be fixed. A good test to check for moisture is to take a piece of plastic and lay it over a small area of the floor, if after a 24 hour period there is moisture on the plastic or the floor is discolored, you may have a moisture problem. Have another professional find the source of the moisture repaired before installation of the floor.

Lay the barrier over the sub-floor. Whether it is a concrete or wood sub-floor, this needs to be installed to assure the new floor will remain free of moisture. The shrinkage and expansion of a hardwood floor and the effect moisture content has on it is illustrated in Figure 4.9.

Moisture content can vary depending on the climate of the area in which the floor is installed. If installed in the Gulf Coast area, moisture content can run from between 11 to 13 percent. In the Western Rockies, moisture content runs 4 to 8 percent. Interior controls such as temperature settings and humidity from showers or un-shaded rooms can affect the moisture content. The moisture content in the hardwood at specific temperatures and humidity is shown in (Figure 4.10).

This knowledge of moisture contents prior to installation of a floor can help in avoiding future problems and insure a long lasting hardwood floor. Once a type of sub-floor has been chosen, it is time to prepare it for installation. All sub-floors must be clean and as smooth as possible. If there are any holes or gaps in the sub-floor they must be filled in with wood filler and sanded down. If there are any points on the sub-floor which are raised, these must be planed or sanded down to a level surface.

INSTALLATION OF HARDWOOD FLOOR PLANKS

Once the type of sub-floor is determined, the floor is ready to be installed. Before installation, a new wood floor needs to be acclimated to the room it is being installed in. This will allow the wood to match the temperature and humidity of the room so that it will not contract or expand while it is being installed. Simply bring the wood into the room being installed for a minimum of 48 to 96 hours, depending on the conditions of the room. It is not necessary

Temperature dry-bulb, °F.	Relative humidity, percent																			
	5	10	15	20	25	30	35	40	45	50	55	60	65	70	75	80	85	90	95	98
30...	1.4	2.6	3.7	4.6	5.5	6.3	7.1	7.9	8.7	9.5	10.4	11.3	12.4	13.5	14.9	16.5	18.5	21.0	24.3	26.9
40...	1.4	2.6	3.7	4.6	5.5	6.3	7.1	7.9	8.7	9.5	10.4	11.3	12.3	13.5	14.9	16.5	18.5	21.0	24.3	26.9
50...	1.4	2.6	3.6	4.6	5.5	6.3	7.1	7.9	8.7	9.5	10.3	11.2	12.3	13.4	14.8	16.4	18.4	20.9	24.3	26.9
60...	1.3	2.5	3.6	4.6	5.4	6.2	7.0	7.8	8.6	9.4	10.2	11.1	12.1	13.3	14.6	16.2	18.2	20.7	24.1	26.8
70...	1.3	2.5	3.5	4.5	5.4	6.2	6.9	7.7	8.5	9.2	10.1	11.0	12.0	13.1	14.4	16.0	17.9	20.5	23.9	26.6
80...	1.3	2.4	3.5	4.4	5.3	6.1	6.8	7.6	8.3	9.1	9.9	10.8	11.7	12.9	14.2	15.7	17.7	20.2	23.6	26.3
90...	1.2	2.3	3.4	4.3	5.1	5.9	6.7	7.4	8.1	8.9	9.7	10.5	11.5	12.6	13.9	15.4	17.3	19.8	23.3	26.0
100...	1.2	2.3	3.3	4.2	5.0	5.8	6.5	7.2	7.9	8.7	9.5	10.3	11.2	12.3	13.6	15.1	17.0	19.5	22.9	25.6

FIGURE 4.9 Shrinkage and expansion caused by moisture content.

Based on Average Possible Change in Width
in Plain Sawn (Tangential Face)

2 - 1/4" Oak Flooring

A Moisture Content Difference of:	May Result in an Approximate Width Change of:
1%	1/128"
2%	1/64" scant
3%	1/64" FULL
4%	1/32" scant
5%	1/32" FULL
6%	3/64" scant
7%	3/64" FULL
8%	1/16" scant
9%	1/16"
10%	1/16" FULL
11%	5/64"
12%	5/64" FULL
13%	3/32"
14%	3/32" FULL
15%	7/64"
16%	7/64" FULL
17%	1/8" scant
18%	1/8" FULL
19%	9/64" scant
20%	9/64" FULL
21%	5/32"
22%	5/32" FULL
23%	11/64" scant
24%	11/64" FULL

FIGURE 4.10 Moisture content of wood at specific temperature and relative humidity.

to lay out each plank separately, but each plank should be exposed to the climate of the room enough to get acclimated.

Next, remove all baseboard trim around the perimeter of the room. Use a crowbar to carefully remove the trim, which can be saved or replaced after the floor has been installed. Be careful when taking off trim so as not to dent or scrape the wall. Place a towel behind the crowbar when using it to remove the baseboard. Also, make any cuts necessary under door jams to allow the hardwood flooring to fit underneath.

NAIL DOWN INSTALLATION

Begin the floor installation along the longest continuous wall. Make a mark ¾ inch all the way around the perimeter of room for the expansion space needed in contraction and expansion. All of this space will be covered once the baseboard trim is replaced at the end of installation (Figure 4.11). Also, the direction of the planks should be at right angles from the joists (Figure 4.12). This will give the best appearance and highest strength for a floor, in general.

When selecting the planks to be used, make sure the finish matches as close as possible to the other planks being used beside it. If there are a few planks that are darker or odd colored, use them in a less visible spot, such as a closet or under furniture or appliances. Face nail the first strip with the grooved side facing the wall (Figure 4.13).

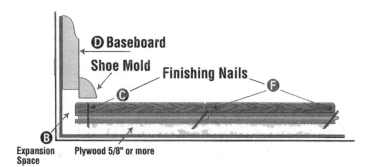

FIGURE 4.11 Nail down application with expansion space under baseboard.

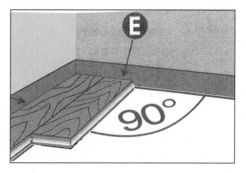

FIGURE 4.12 Installation of wood floor at a 90 degree angle.

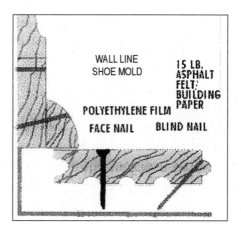

FIGURE 4.13 Face nail the first strip of wood.

Place the row of nails as close to the wall as possible. Once you have moved a little further away from the wall, a power nailer can be used.

Once the first plank is nailed down, cut the second strip at least 6 inches shorter than the first plank. The plank ends should not line up with one another. This will give the floor a more natural appearance and less chance of stress on the ends. All wood planks are made with the tongue-and-groove construction. So, simply fit the tongue into the groove and nail down the exposed tongue of the strip. Use 1½ inch nails every 4 inches along distance of the plank.

fastfacts

When working with a floor that has a width of more than 20 feet, special consideration should be made. Rather than starting on the longest wall, begin at the center of the room. Nail down a starter strip groove and glue in a slip tongue. Install the floor in opposite directions, adding additional expansion spaces as needed. This will give the floor more space to move and avoid cupping and warping later on.

On the ends, do not place nails within 2 inches of one another, this may cause splitting of the wood. Be sure to nail down the ends of each strip with at least two nails. Continue this same process throughout the installation.

When you come to any cutouts in the floor, make a template of the shape and cut it out of the plank with a hand saw or router (Figure 4.14).

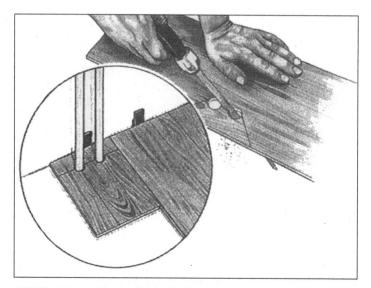

FIGURE 4.14 Make a template of a cutout.

fastfacts

Use nails no longer than 1½ inch when nailing into a concrete slab sub-floor covered with plywood. After the third run of nails are in place, you may switch to a power nailer. This machine allows the nail to be counter sunk at the same time so no follow up is required with a nail punch. When using the power nailer, a 1½ inch cleat is recommended rather than the standard 2 inch. The longer cleat may come out the bottom side of the plywood and tear the vapor barrier underneath.

When you reach the last row of planks and you have to make an uneven cut, take a piece of wood and lay it over another piece face down. Take a chalk line and mark along the length of the piece of wood. This will give you the angle for cutting the last strip. Once cut, hand nail into place along length of wall, and replace any transitions or baseboard around the perimeter of room (Figure 4.15).

GLUE DOWN INSTALLATION OF HARDWOOD FLOOR

As with the nail down method of hardwood flooring, you must find the starting line. Measure out 31 inches from the longest wall and mark with a chalk line. Take a piece of 1 x 2 x 8 feet pine and nail it into place along the chalk line. This wood will hold your first row of starter boards in place. Follow the manufacturer's instructions in applying adhesive directly to sub-floor.

Drying time of adhesive can vary depending on temperature and humidity. Do not apply adhesive to an area you cannot finish in less than two hours. Apply the adhesive with a trowel at a 45 degree angle. Make it a thin layer using the ridge side of the trowel (Figure 4.16). Too much adhesive can make the floor uneven and cause movement as you install the remaining planks.

Place the first plank down, tongue side facing the wall. Always work from left to right when laying the planks. Continue setting the planks using the tongue and groove method. Gently tap the planks together with a rubber mallet and then press down into the adhesive.

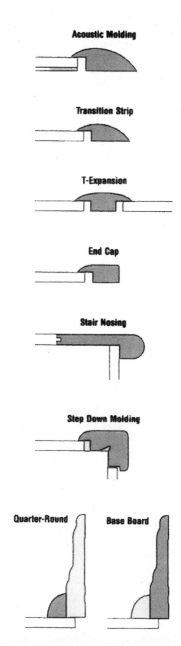

FIGURE 4.15 Replace baseboard molding.

FIGURE 4.16   Apply adhesive to starter strips.

Lay full planks until cut pieces are needed to finish a row. Be sure to leave ¾ inch gap at the end of the row along the wall for expansion. Continue this same installation process, never allowing any ends to match one another. When you reach the last row of planks, make sure they are even with the wall. If not, you may need to cut an uneven strip to finish the last row. Place a piece of flooring down and lay another face down on top along the uneven wall. Make a chalk line to find the correct angle to cut the strip.

Once the strip is cut, hand nail it into place to finish installation. If there are any cutouts in the floor, make a template of the shape and cut it out of the plank with a hand saw or router. After the floor is completed, replace any transitions and baseboard around perimeter of room. Do not let any adhesive get into the grooves of the strips, this may not allow the rows of planks to fit tightly together. If any adhesive gets on the hardwood floor surface, wipe it off with a wet towel. If adhesive is dry before it is wiped off, use a solvent or mineral spirits.

HARDWOOD PARQUET FLOOR INSTALLATION

Parquet floors can be installed two different ways, either square with the walls of a room or at a 45 degree angle. Both must be laid over a sub-floor of mastic, spread at a rate of 35 to 40 feet per gallon. Let the mastic dry over night so it is dry enough to make a chalk line over it.

SQUARE PATTERN LAYOUT

When laying out a square pattern, do not use the walls as your guide. Most walls are not perfectly square, so measure out about 3 to 4 feet from the door of the room. Make a chalk line five parquet units away from the door. Next, find the center point of this base line and snap another line at a 90 degree angle to it from the wall. Use this line as your guide for the entire project to keep it square (Figure 4.17).

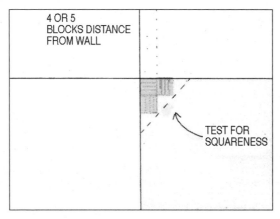

4 OR 5
BLOCKS DISTANCE
FROM WALL

TEST FOR
SQUARENESS

FIGURE 4.17   Chalk lines for square parquet layout.

fastfacts

A good test for squareness is to measure four feet along one line from where they intersect, and three feet along the other. The distance between these two points should be five feet.

Diagonal Pattern Layout

Measure equal distances from one corner of a room, along both walls, and snap a chalk line between these two points to form the base line. A test line should intersect the center of the base line at an exact 90 degree angle (Figure 4.18).

Most existing parquet pattern can be laid out using these two working lines. A herringbone pattern will require two test lines, one at a 90 degree angle; the other crossing the same intersection at a 45 degree angle to both.

For any wood parquet, always install in a pyramid or stair-step pattern, rather than in straight rows. This allows a more natural appearance and hides any discrepancies in the parquet. Place the first parquet at the intersection of the base and test lines. Lay the next squares ahead and to the right of the first one, following the chalk line. Continue installation in the stair-step method, making sure the corners are aligning with one another.

Install a quadrant at a time, returning to the base and test lines each time you start a quadrant until all are completed. Install the last quadrant from the base line to the door (Figure 4.19). Cut any remaining tiles for the perimeter of the room leaving an expansion space of ¾ inches.

When installing wood parquet in a corridor or a room where the length is more than 1½ times the width, a diagonal pattern is recommended. This layout minimizes expansion under high humidity rates. Cut out any openings such as pipes or vents by making a template of the area. Then use a hand saw or router to cut the shape out of tile. Once all the tiles are completely installed, replace any transitions or baseboard around perimeter of the room.

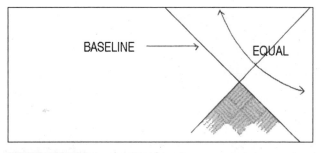

FIGURE 4.18 Diagonal pattern layouts.

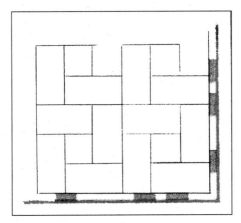

FIGURE 4.19   Quadrant floor layout.

fastfacts

Most mastic allows hardwoods to slip when sideways pressure is applied. Avoid this sideways pressure by laying plywood panels on top of the installed area of flooring. Also, do not replace furniture or allow activity on the floor for a minimum of 24 hours. In extreme cases, rolling the floor may be necessary to stabilize any possible movement.

SPECIALTY OR INLAID WOOD FLOORS

The installation of an inlaid floor is a job for a highly experienced professional. It is a precision job which can dramatically increase the appearance of a floor. You can install an entire floor with inlays or just use it as a border design. Any inlay can use a variety of exotic woods such as Brazilian Cherry, Ebony, Curly Maple, the list goes on and on (Figure 4.20).

It is a very costly and time consuming process, but makes a floor unique from all others. Inlays placed into an existing floor are cut

FIGURE 4.20   Examples of wood inlay floors.

out of the wood floor with a precision router at a depth of ⁵⁄₁₆ inch. The inlay is made from a pattern of the cut out space, and then glued back into the pocket created by the router. Never face nail an inlay into an existing floor, always use glue.

FINISHING AN UNSTAINED WOOD FLOOR

Most hardwood floors come pre-finished with a stain and several coats of varnish. If you are installing an unfinished wood floor, it needs to be treated before use. Test any stain on a scrap piece of the wood floor first before applying to the actual floor. Once the correct stain is chosen, follow manufacturers instructions for application. After the stain is applied, a protectant must be applied over it. There are two different types of varnish available, water-based varnish and polyurethane varnish. Water-based products are quick-drying, low odor, non-yellowing, and clean up with water. Polyurethane products can take up to 24 hours to dry, have a strong smell, require chemicals for clean up, and can yellow over time. Polyurethane varnish is better for harder wear-and-tear. Remember that all types of varnishes can darken the surface of any hardwood.

Hardwood floors are meant to last a lifetime. The proper installation will play a large part in how long that will be. Use the right tools and make sure to protect against moisture and humidity by using vapor barriers and correct sub-floors. Follow all the manufacturers' instructions carefully for adhesives and mastic. This can help insure the longevity of a floor and avoid any future problems like warping and cupping. Be highly accurate when laying out patterns and planks to leave room for expansion. By following all these basic guidelines, installation can be quick and easy.

chapter 5

CERAMIC TILE
INSTALLATION

Ceramic tile offers a huge variety of choices for different types of projects. A vast array of colors, shapes and sizes are available which can be used in conjunction with complimentary wall tile or as a trim option.

Tile floors are the most expensive to install and very time consuming. Many other types of flooring can be installed with an existing sub-floor or have plywood as the underlay. This is not always the case for ceramic tile. A choice of up to four different sub-floors has to be decided upon. After the ceramic is installed, a grout must also be applied between each tile and if the tiles are not already sealed, this too must be applied. All of these options add time and cost to this type of flooring installation.

Tile is the heaviest of all flooring materials, especially when adding the sub-floor which could be a cement backer board. A supporting floor must be extremely well braced and perfectly level before installation of any type of ceramic or natural stone tile and its sub-floor.

When choosing tile, beware of trendy styles and colors. A proper installation of ceramic tile will last for decades. Pick a color which will have a long lasting appeal and work with future design trends.

TYPES OF TILES

There are two types of tiles available in individual pieces or in sheets. Ceramic tile is the most common form of floor tile and is the least expensive of the two. Natural stone tile is more expensive and includes granite and marble flooring. Since these tiles are not man-made, there are greater variations in the finishes and sizes.

Ceramic Tile

Ceramic tile is available in many different forms (Figure 5.1). These are tiles made of baked clay then fired in a kiln. After it is taken out of the kiln, a coat of color glaze is applied and then baked in the kiln again to set the finish permanently.

Quarry tile (Figure 5.2) is an unglazed tile which is a little thicker than the glazed ceramic. This type of tile is also softer and more porous than glazed tile.

Porcelain mosaic tile is a hard, dense tile which is naturally water-resistant. This type of tile is usually manufactured in sheets of small squares with a paper backing applied to the back (Figure 5.3).

FIGURE 5.1 Ceramic tile.

FIGURE 5.2 Quarry tile.

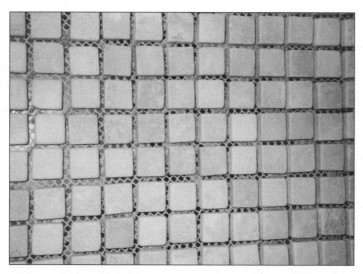

FIGURE 5.3 Porcelain mosaic.

Natural Stone Tiles

Natural stone tiles are exactly what they sound like, tiles cut from stone. These tiles are cut from quarries around the world, so there will be many variations. Marble and granite slabs are available in polished and sealed pieces cut to standard or specific sizes. This type of stone tile is smooth and shows all the veins and spots nature gave them. Slate is another option of natural stone tiles (Figure 5.4). This tile has a lot of texture and unevenness for its surface.

All of the natural stone tiles are extremely expensive if used for a large project. A mix of natural and man-made tiles can be used to cut down on cost and give the installation a unique appearance. Natural stone tiles are great for accent in more visible areas such as an entryway.

SIZES OF TILE

Square tile is the most common shape manufactured and sold on the market. Sizes of the squares range from 6 to 12 inches, but lately the 14 inch square tile has been gaining in popularity. The larger a tile is, the quicker it can be installed and less product is necessary, saving

FIGURE 5.4   Slate tile.

time and money. There are irregular tile shapes available such as octagons, diamonds, hexagons and smaller squares. These smaller tiles are usually sold in sheets of 1 or 2 inch squares, making installation and alignment easier. Smaller pieces of tile are lined up and a paper or fiber backing is applied to them.

Accent tiles are typically horizontal in shape and can be any size or shape (Figure 5.5). These types of tiles are made to fit in with most any size tile as a border or to accentuate other tile being used.

TILE COLORS

The color of tile is dependent on it being man-made or a natural stone tile. Man-made ceramic tile can be purchased in hundreds of standard colors or a custom glaze can be produced to match just about anything.

Natural stone tiles are pretty much set in stone. Whatever is quarried from the earth after millions of years underground is what is available. These pieces are extracted raw from the ground and can be porous and unpolished. After the cutting, polishing, and sealing process are finished, the original finish and texture is enhanced, but the color is basically unchanged.

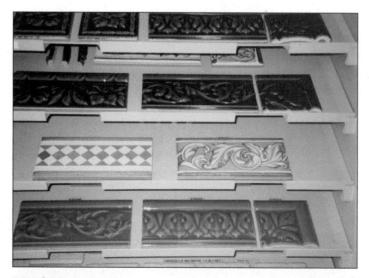

FIGURE 5.5 Accent tiles.

MEASURING THE FLOOR FOR TILE INSTALLATION

Once the type of tile is decided upon, the floor must be measured for product. Square tiles are sold individually or in a box which covers a specific area. The coverage area per box is listed on the side, along with the color. To measure, take the room length and multiply that by the room width to get the total square footage. Remember to calculate an additional 5 to 10 percent for mistakes and special cuts. A handy chart is shown below for use with 12 inch square tiles (Figure 5.6).

TOOLS NEEDED FOR CERAMIC TILE INSTALLATION

Here are the tools needed for installing ceramic tile. These include tools and materials to complete the job, from product to clean-up.

Tool Checklist

✔ Notched trowels—Notch size should be according to size of tile being installed. These are used to spread the adhesive and grout solutions
✔ Tile cutter
✔ Tile nippers
✔ Hand-held tile cutter
✔ Jig saw with carbide blade
✔ Rubber mallet
✔ Grout float
✔ Buff rags
✔ Tile spacers—For use between larger size tiles
✔ Caulk gun
✔ Caulk—Use instead of grout when tile meets another surface
✔ Level
✔ Chalk line
✔ Tape measure
✔ Framing square
✔ Tile file

ROOM LENGTH (ft)

ROOM WIDTH (ft)	6'	7'	8'	9'	10'	11'	12'	13'	14'	15'	16'	17'	18'	19'	20'
6'	1	1	2	2	2	2	2	2	2	3	3	3	3	3	3
7'	1	2	2	2	2	2	2	3	3	3	3	3	3	4	4
8'	2	2	2	2	2	2	3	3	3	3	3	3	4	4	4
9'	2	2	2	2	3	3	3	3	3	3	4	4	4	4	5
10'	2	2	2	3	3	3	3	3	4	4	4	4	5	5	5
11'	2	2	2	3	3	3	3	4	4	4	4	5	5	5	6
12'	2	2	3	3	3	4	4	4	4	5	5	5	5	6	6
13'	2	3	3	3	3	4	4	4	5	5	5	6	6	6	6
14'	2	3	3	3	4	4	4	5	5	5	6	6	6	7	7
15'	3	3	3	3	4	4	5	5	5	6	6	6	7	7	7
16'	3	3	3	4	4	4	5	5	6	6	6	7	7	7	8

FIGURE 5.6 Room calculation chart for 12 inch square tiles.

All of these tools should be ready to use at any time during the project. When it comes to sub-floors, additional supplies may be necessary. These options will be discussed in the sub-floor section of this chapter.

PREPARING THE SUB-FLOOR

In this section we will go over different types of sub-floors you can use for ceramic and natural stone tile. Here are ways to prepare the particular sub-floor you are installing tile on. Remember, tile is the heaviest product you can install for flooring, so the floor has to be able to support it.

If you are starting with a concrete floor, you need to make sure it is absolutely level. If there are any gaps or holes, they must be filled in and left to dry before adhesive is applied. A tile floor can be installed right over a perfect concrete floor. If the concrete surface is lightly roughed up, adhesion is easier to establish.

When installing tile over an existing wood floor, make sure to lay AC grade plywood or cement backer board over it. Installation of plywood is the same as with any other type of sub-flooring. Make sure the joints do not meet and that all nails are set and filled in to achieve a level surface.

UNDERLAY MATERIALS FOR TILE INSTALLATION

There are basically five different underlay materials than can be used individually or in conjunction with one another.

- *Plywood*
- *Fiber cement board*
- *Cement board*
- *Isolation membrane*
- *Waterproof membrane*

Let's go over the list of these products and discuss what each one is used for. Plywood is the most common underlay for a variety of floors. This is the least expensive and quickest way to put down a sub-floor. A layer of ½ inch AC plywood is recommended under ceramic tile. If there is a wood-based product being covered by a new sub-floor, plywood is the way to go.

Fiber cement board is a dense underlay that is used under ceramic tile basically for thickness. If you are trying to build up a floor that is going to meet a different type of floor, such as carpet with padding, fiber cement board is a good choice. This height difference can also be achieved with two layers of plywood as your sub-floor rather than one. While plywood is the less expensive option, it will take twice as long to install.

Cement backer board is specifically designed for ceramic tile installations (Figure 5.7). This product remains completely stable, even when wet, so it works best in areas that can get wet or attract moisture. Bathrooms, laundry rooms and above grade basements are good places to use this type of underlay.

A good test to see if cement backer board is needed would be to walk across the floor. If there is any movement then a layer of cement board is necessary. Cement backer board is available in ½ inch and ¼ inch thickness. When installing backer board over a wood sub-floor, make sure the total thickness of both plywood and backer board is a minimum of 1 inch. Measure the area being covered and cut the board accordingly. Use a straight-edge and scoring tool to make the cuts on the backer board. Snap off the board at the scored line and cut the mesh on the back side with a wire cutter (Figure 5.8).

If there are cutouts or transition, use a drill or circular saw to make the cuts, then knock out with a hammer (Figure 5.9).

FIGURE 5.7 Cement backer board.

FIGURE 5.8 Scoring and cutting cement backer board.

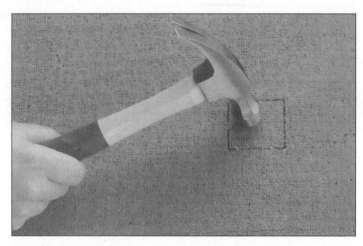

FIGURE 5.9 Cutout for cement backer board.

If applying the backer board over an existing linoleum or vinyl floor, measure and cut the same way. With this sub-floor, you must apply a thin coat of mortar directly on to the existing floor before laying the backer board. Use a ¼ x ¼ x ¼ inch square-notched trowel when applying the latex thin-set mortar (Figure 5.10). Always follow the manufacturers' instructions on the label when applying mortar.

When laying the backer board over plywood, place the sheets perpendicular to the plywood. Leave a ⅛ inch gap between each sheet of backer board and stagger the corners, just like a plywood sub-floor installation.

Never apply cement backer board over a cushioned or sponge-backed floor. A cushioned floor should be taken up first and a new sub-floor will have to be installed. After the board is installed, fasten it down with 1¼ inch corrosion-resistant backer board screws or 1½ inch galvanized roofing nails. Around the edges of each board, fasten every 6 to 8 inches and within ½ to 2 inches throughout each panel.

Fill all the gaps between the boards with thin-set mortar, leaving ⅛ to ¼ inch gap for expansion around the perimeter. Taping the joints is not required on floor or counter joints, only when installed on a wall. Cement backer board is probably the most expensive product you can use, but it pays for itself if there is even the slightest chance of future moisture problems.

Another product that can be used for sub-floors is isolation membrane. This protects a floor from movement that can occur when the floor settles or has existing cracks (Figure 5.11). It can cover individual

FIGURE 5.10   Application of mortar under cement backer board.

cracks or an entire floor if necessary. This highly specialized product is not used in most installations.

Waterproof membrane can stop the moisture from coming up through cracks. This should be used in all shower installations, placing it over the sub-floor and under cement backer board.

FLOOR LAYOUT

Before the actual installation of the tile, make a chalk grid on the floor to see exactly where each tile will go. This can assure that you have enough tiles to complete the project. Find the longest wall in the room, then measure to find the center of the wall. From that center point, snap a chalk line across the room at a 90 degree angle to the wall. On the right side of the center line, dry lay a row of tiles using recommended spacing between them (Figure 5.12).

Proper spacing is determined by the size of the tile. You can adjust any joint spacing per installation, keeping in mind the larger the space between tiles the more likely they are to crack. Irregular size tiles usually require larger spaces between them.

Glazed tiles—³⁄₁₆ to ⅜ inch space

Porcelain tiles—⅛ to ¼ inch space

Terra-cotta tiles—¾ inch space

Cement-bodied tiles—⅜ to ½ inch space

Natural stone tiles—Up to ⅛ inch space

FIGURE 5.11 Isolation membrane.

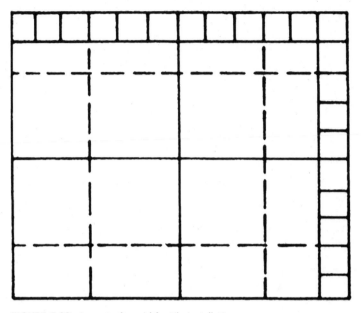

FIGURE 5.12 Layout of a grid for tile installation.

fastfacts

If the tile in the corner is less than half a tile wide, erase the chalk line or use a different color chalk, and snap a new center line half a tile width to the left. This will give you evenly sized tiles at the end of the row. Then go to the adjacent wall and repeat the same process.

To complete the grid layout, go over three tiles plus three grout joint widths from the first center line and make a chalk line. Next, take your metal square and line up one side with the second center line. From the edge of the third tile make another chalk line across the room at a 90 degree angle to the wall. Repeat this process all the way down the longest wall, then the adjacent wall. When complete, your floor should look like a one large grid.

ADHESIVES

All ceramic tiles can be adhered to most types of sub-floors with a variety of adhesives. In the past, a thick bed of portland cement was used, but new adhesives have been introduced that are just as solid and level for setting of tile.

The most common type of adhesive is thin-set. This combination of sand and cement in a powder form is available in 5 to 50 pound bags. Just add water or a latex additive for more flexibility to create a thick adhesive and a strong bond for the tile.

Epoxy adhesives are another option, but are more costly and harder to apply. They do provide a far stronger bonding and greater resistance to impact, water and chemicals. Epoxy can adhere to any surface, including existing tile and metal. This type of adhesive comes in a two-part package and, when mixed together, become toxic to the skin. Be sure to follow manufacturers' instructions carefully when using.

Organic mastics are the third option that can be used to adhere tile to a sub-floor. These can be latex or solvent-based organic mastics ready to use in a premixed paste form. The basic difference is the latex is easier to clean up and not flammable, like the solvent-based mastic.

Organic mastics are primarily used on plywood or drywall sub-floor and are extremely easy to apply. They do not bond as well as the other adhesives and are not recommended for wet areas such as a bathroom or laundry room. When applying any tile adhesive the proper trowel must be used.

- *Mosaics/small tile use a ³⁄₁₆ x ⁵⁄₃₂ inch or ¼ x ¹⁄₁₆ inch, v-shape*
- *Flat-backed tiles use a ¼ x ¼ inch, square shape*
- *Irregular/lug-backed tiles use ¼ x ⅜ inch or ½ x ½ inch, square shape*
- *Marble/granite use ¼ x ¼ inch or ¼ x ⅜ inch, square shape*

APPLICATION OF ADHESIVE

Mix adhesive according to the manufacturer's instructions. When the adhesive has a consistency of frosting it is properly mixed. Let the mix set for about ten minutes before application to the sub-floor. Start the application in the grid farthest from the door. Choose the right size trowel, using the flat side to spread the adhesive in the grid area. Never apply more to an area than you can work with at one time. Now take the notched side of the trowel and comb the adhesive at a 45 degree angle to the floor (Figure 5.13).

Now take a full tile and place a small amount of adhesive on the back for a stronger bond to the sub-floor. Position the first tile at the intersection of the horizontal and vertical chalk lines. Press the tile down into the adhesive and tap down gently with a rubber mallet. Place spacers between each tile for consistency, according to the size and style of tile (Figure 5.14).

If the tiles you are using have tiny lug spacers along the four edges, there is no need to use additional spacers. The lug spacers are placed on some types of tile for a more consistent grout line. Continue this same process until completing one grid area. Take a two by four and move it across the tiled area, checking for any high or low spots. If there are any differences tap them down with a beater block or take up the tile and add a bit more adhesive to raise the tile. Complete each grid area until all the areas are covered with full size tiles.

BASIC TILE CUTTING TOOLS AND TECHNIQUES

When cutting a tile is necessary, a few tools can be used to make the proper cuts. For simple cuts, a manual tile cutter can be used to score the tile, and then snap it apart. When making a more difficult

FIGURE 5.13 Apply adhesive with notched trowel at a 45 degree angle to floor.

FIGURE 5.14 Tile spacers.

cut or several cuts, a wet saw can make the job quick and easy. This type of saw has a diamond blade which makes the cut faster (Figure 5.15) A wet saw can also make chips in a tile as it cuts which can usually be covered with grout when finishing the floor.

A heavy duty set of tile-nippers can be used to bite away at a tile until the desired size or shape is attained. This tool works best for small jobs and removing small bits at a time.

A tile file is used after a cut has been made to tile. This is a sturdy sanding tool that gets rid of the rough edges around a tile.

CUTOUTS AND TRANSITIONS

When you need to cut away at a piece of tile, any of the above tools will work. Cutouts can be pipes, vent openings or the corners of a room.

Remember, when you lay new sub-floor, the addition of adhesive and tile your floor will have raises the floor a minimum of one inch. So be sure to cut away under any door frames to make room for the new floor height (Figure 5.16).

FIGURE 5.15 Wet saw cutting a tile.

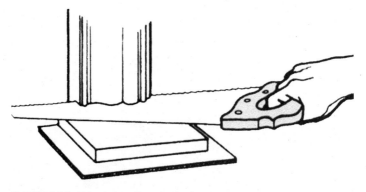

FIGURE 5.16 Wood door frame cutout.

To find out space needed, simply place a tile on a thin piece of cardboard next to bottom of door frame. Take a handsaw and lay it on top of the tile and cut into the bottom of the door frame. When it comes time to tile under the frame, spread adhesive under the tile and slip it under the frame.

Be sure to leave a ¼ inch expansion gap around the perimeter of the room. Check the tile for any excess adhesive, making sure to wipe any off with a damp cloth before it dries.

INSTALLING STEP TILES

When tiling steps, be sure to install the front side of the step first. Measure and cut the tile just as you would a regular floor tile. Adhere and space the tile the same as floor tile. Then install the top tile of each step, slightly overlapping the exposed top of the front tile.

INSTALLING THE GROUT OVER THE TILE FLOOR

After the tile floor has set and cured for at least 24 hours, it is ready for grout. Before grout is spread over the floor, remove all spacers from between the tiles using needle nose pliers. Hopefully you were able to check the floor while it was still wet for levelness. Once the floor is dry, it is very difficult to remove and replace a tile that is uneven.

fastfacts

When a drill is needed to make a hole through a tile, great care must be taken not to crack or splinter the tile. Use a carbide-tipped blade or a masonry bit. To protect the tile when drilling through it, place a ball of plumbers putty on the spot where the bit will go through. Place a drop of water directly in the center of the putty ball. The wet putty will act as shield, protecting the rest of the tile while you drill.

Next, replace any transitions that were removed between the rooms. When choosing grout, there are a few options. You can choose from a cement-based grout made with a mixture of portland cement, sand and additives. The other option is latex-portland cement grout made up the same way as cement-based grout only with a latex additive.

Grout is available in a variety of basic colors which will complement most ceramic and natural-stone tile. There is a hardened polymer-modified grout which will make the color grout last longer and more water-resistant. Always follow the manufacturers' instructions for mixing and application of grout.

Mix the grout to a consistency of mashed potatoes. Once the mix has set for ten minutes, it is ready to spread over the tiles. Do not work an area larger than 3 to 4 feet at a time. Spread the grout over the floor and use the rubber grout float to work it into the joints. Hold the grout float at a 45 degree angle and smooth away any excess grout (Figure 5.17). Be sure to clean the float often for optimal spreading

fastfacts

To protect the walls around the base of the perimeter, mask the wall with wide tape so the grout does not stain it. Also, wipe grout release over the tiles before application of the grout. This will help to easily remove any extra dried grout from the tile surface during clean-up.

FIGURE 5.17 Applying the grout.

conditions. Continue to work the grout into the joints until all has been smoothed off the tile and into the joints. Work at the floor in different directions, making sure to fill in all joints completely with grout.

Wipe off any excess grout with a damp sponge. Remember that grout is weakened by water, so wring out the sponge or cloth if it contains too much water. Wipe out any grout in the expansion joints around the perimeter.

After 20 to 30 minutes a dry haze may appear on the tile. Wipe the tiles off with a dry, non-abrasive cloth until the haze has disappeared. After the grout has cured for several days or weeks, depending on the climate, apply a silicone or acrylic sealer to the grout lines. This will help keep the dirt out of the grout lines. If tile was installed in a bathroom or laundry room where moisture could get under the tile, apply a thin line of caulk between the wall and tile.

The installation of a tile floor can be time consuming and costly. The most important factor to remember is that the sub-floor must be strong enough to support the heavy weight of a tile floor, plus the extra layers of underlay material and sub-floor. Proper measurement and practice in cutting tiles can also make the job quick and easy. More elaborate designs can be installed as experience is gained on the job. Tile floors can add a unique look to any floor while adding to the value of a home.

6

RESILIENT VINYL FLOORS

Resilient vinyl floors represent the most in color, style, and ease of installation. There is no better value for the dollar than installing vinyl flooring. Whether choosing sheet flooring or square tiles, the durability factor of vinyl is hard to deny. Resilient flooring holds up under the most extreme conditions and offers the greatest resistance to stains. This type of flooring also provides great noise abatement qualities and the most comfortable feel of any hard floor surface. When it comes to maintenance, no other floor is easier to clean and more resistant to wear and tear. If maintained properly, this type of flooring will last for several years.

When choosing a style and color for this type of floor, be sure to pick the right product for the right room. Resilient flooring is an excellent choice for bathrooms, kitchens and laundry rooms. It is extremely water resistant and holds up to the wear-and-tear of kids and pets. Try to pick a style that will work with the current color scheme or theme in a room. This floor should last a long time, so stay away from trendy styles and colors.

Colors and patterns also play a part in making a room appear larger or smaller. For example, if a room is large-scale, choose a small-scale design and vice versa for a small-scale room. This will create a look where no two patterns overwhelm the room. A lighter color tile will help make a room appear larger. Dark tones will absorb light and create a more ambient look. Also try to choose the color in the same light as the room in which it will be installed. Take a tile or a piece of the sheet and place it in the room during daylight and nighttime lighted conditions.

VINYL CONSTRUCTION

Resilient floors are available in two forms, tiles or rolled sheets. The sheet vinyl is available on rolls between 6 and 12 feet wide. The tiles are available in 12 inch squares, either with a peel back adhesive or ready to install after adhesive is applied to a sub-floor. Vinyl tiles tend to be the more resilient of the two, having more give and expansion room than the sheet vinyl. Resilient floors contain some high tech plastics not available back in the days of linoleum. The same plastics which are used in plumbing and sewer pipes are now used to manufacture floors. Because of the durability factor, polyvinyl chloride, PVC, is used in most resilient flooring today.

The part of the vinyl floor which is most important is the top layer. This is also called the wear layer. PVC is used in the manufacturing of the wear layer, while adding durability to the floor; it also adds the ability to seam different sheets together. When using special cements which join seams together, the PVC molecules will interact between the two sheets and the cement actually creating one piece of flooring.

Wear layers are available in different thicknesses. The thickest wear layer is 0.025 inch and the thinnest is 0.005 inch thick. The thin wear layers are the most susceptible to damage, but are the least expensive. This is important to consider when choosing the floor. If the pattern or colors become damaged, it cannot be repaired easily. Vinyl resilient floors are available in different grades, based on thickness and wearability:

- *Commercial Grade—This grade of vinyl is primarily used in commercial buildings where high traffic is prevalent. It is solid vinyl throughout the entire piece. Most commercial grade vinyl requires pre-installation techniques, i.e. heat welding of seams.*

- *Premium Residential Grade—This type of vinyl is made of three layers with a thickness of 3/32 inches. The bottom layer is a felt backing, the middle layer is a foam cushion and the top layer is vinyl. Although not the top of the line, it offers excellent wearability. (Figure 6.1).*

- *Standard Residential Grade—This vinyl is made the same way as premium grade, but is slightly thinner. It offers good wearability. (Figure 6.2).*

• *Spec-Grade—This vinyl is a single-layer product ¹⁄₁₆ inch in thickness. It is the least expensive and offers fair wearability. (Figure 6.3).*

Choose a floor with the highest grade and it could save time and money in the future. If a low grade is used, replacing all or part of the floor before its time would not be economical or time conscious. If a high gloss finish is required for the floor, both tiles and sheet vinyl can have several coats of urethane applied at the factory. Remember, these high gloss finishes will show more wear and tear, requiring more maintenance.

INSTALLATION TOOLS AND MATERIALS

Here is a list of tools and materials needed to install vinyl sheet flooring. Depending on the sub-floor being used, not all tools may be necessary.

Tool Checklist

✔ Straight-blade utility knife and extra blades
✔ Hammer
✔ Short nap paint roller
✔ Rolling pin
✔ Scissors
✔ Flat screwdriver or putty knife
✔ Carpenter's saw
✔ Smooth edge trowel for spreading adhesive
✔ Metal straightedge
✔ Medium grit sandpaper
✔ Tape measure
✔ Chalk and chalk line
✔ Carpenter's square
✔ Glas-tac acrylic double-sided tape
✔ One part embossing floor leveler
✔ Adhesive
✔ Patch and skim coat

FIGURE 6.1 Premium residential grade vinyl.

FIGURE 6.2 Standard grade vinyl.

FIGURE 6.3 Spec-grade vinyl.

fastfacts

If installing a vinyl floor off a driveway or garage entrance which is paved with asphalt, this will interact with the PVC wear layer. Small amounts of oil are picked up by shoes from the asphalt and can be deposited onto the PVC wear layer. This oil will permanently stain the vinyl floor and a distinct yellow pattern will develop over the traffic path. Area rugs or the removal of shoes are highly recommended if this type of floor is chosen for this area.

Gather all the proper tools and materials before starting the project. Remember to always follow the manufacturer's directions on the product label for best results.

MEASUREMENTS FOR SHEET VINYL AND VINYL TILE

Once the floor type has been chosen, it is necessary to measure the room. A drawing of the room is a good idea so you can make marks where doorways and cutouts will be. Also, if the floor is larger than the width of the sheet vinyl, a seam location will have to be determined. Be aware of the pattern of the floor, keeping it centered and not covered by cabinets or placed along the edge of a wall.

SHEET VINYL MEASUREMENTS

Vinyl flooring comes in 6 or 12 foot widths cut to the length needed. Find the total square feet of the room by multiplying the length by the width. This will give the total square feet.

Since sheet vinyl is sold by the square yard, take the total square feet and divide that number by nine (Figure 6.4). Be sure to add a minimum of six inches all the way around the perimeter allowing for cutting and mistakes. This formula is used if there is no seam needed.

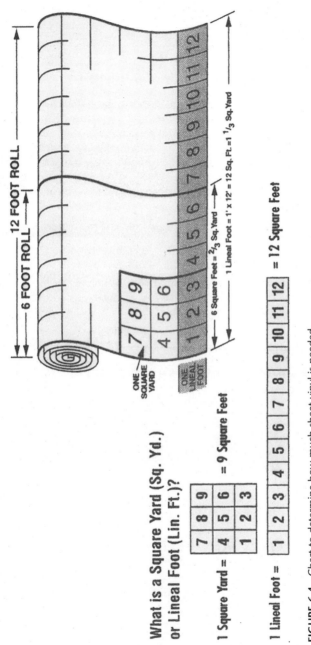

FIGURE 6.4 Chart to determine how much sheet vinyl is needed.

A seam is needed if either the width or length of a room is more than 12 feet. Lay out one length of a 12 foot wide sheet that will give the most coverage to the area. Be sure to leave the additional six inches for cutting around the perimeter of the room. If the piece gives the pattern a centered look, take a second piece and match up the patterns. On the second piece, do not add the additional six inches. Measure the actual repeat of the pattern and add that to the section of floor that remains to be cut. This will give the actual amount needed to complete the floor.

Vinyl Tile Measurements

Determine the total square feet of the room by measuring length by the width. Vinyl tile is available in 9 or 12 inch squares. To calculate how many 12 inch squares are needed, determine the length times width for total square footage. The total square footage is exactly how many tiles are needed for a project. Nine inch tiles can be calculated by multiplying the square footage by 1.78. That result will show how many tiles are required.

SUB-FLOOR OPTIONS AND PREPARATIONS

The sub-floor you choose to install under the floor is an important decision. It all depends on what is already in place or if it is new construction. For either the sheet vinyl or the vinyl tiles, resilient vinyl product can be installed over existing vinyl, plywood, or concrete and ceramic tile.

INSTALLATION OVER EXISTING TILE

If you are considering installing new vinyl tile over existing vinyl, be sure to check for these important factors. The old vinyl must have a non-cushioned back. There must be no loose or torn vinyl sticking up. Be sure to check for large gaps around the perimeter and seams. These will need to be filled before any new floor is installed.

For tears or cuts, remove the excess with a utility knife and level out with one-part embossing leveler. Do the same with any large gaps, wider than ⅛ of an inch. After all of those repairs have been made and the leveler has dried, remove the dirt and wax from the existing floor. Any brand of floor stripper and cleaner is fine to take the shiny coat up and help the new floor establish a hold.

INSTALLATION OF PLYWOOD SUB-FLOOR

If the floor is beyond repair or the existing plywood sub-floor has been damaged, install a new plywood sub-floor. Use APA Underlayment Grade plywood under either resilient sheet or tile floors. Cover the entire surface making sure that the plywood corners are staggered. Leave an expansion gap around the perimeter of the room of ⅛ inch or less when installing the sub-floor. Use coated or ring-shanked nails, making sure nail heads are even with floor. For both existing and new plywood, fill in any holes, including all nail holes and gaps between the planks. Do not fill in the expansion space around perimeter of room. After any filler has completely dried, use medium grit sandpaper and go over the floor to smooth out any rough surfaces. Once everything has dried and been properly sanded, apply a coat of latex primer over the entire surface of the sub-floor. When dry, the primer will create an excellent bond between the sub-floor and the new tile.

INSTALLATION OVER CONCRETE OR CERAMIC TILE

A concrete sub-floor can be a great surface on which to install vinyl tile. Make sure the floor is dry, clean and dust-free. If there has been any moisture on the floor, make sure that it has been repaired. If there is still any chance that the floor will get wet again, do not install the vinyl tile over the concrete. Lay a moisture barrier first or the vinyl will warp and have to be replaced.

Check for cracks and levelness of the floor. Use fast setting cement or patch to repair any problems. Follow the manufacturer's directions for best results. Once leveler and patches are dry, go over the floor with medium grit sandpaper, smoothing out any rough areas. Apply a coat of latex primer to create a better bond between the concrete and new floor. Self-stick tiles can be installed over painted concrete floors.

New vinyl flooring can be installed over existing ceramic tile, terrazzo and marble. Make sure the old floor tiles are securely bonded before placing the new tiles over them. Check for any large cracks or gaps and repair with patch or leveler. Apply a coat of latex primer to assure adhesion of the new floor surface.

PRE-INSTALLATION PREPARATION OF THE VINYL FLOOR

First, when storing or moving the floor product from warehouse to site, make sure not to bend or kink. This can lead to permanent distorting of the tile or roll. Sheet vinyl should be rolled face side out until ready to install.

When possible, lay out the whole roll flat prior to installation. Both the floor and room temperature should be between 65 and 85 degrees for 48 hours before the installation as well as during and after the installation.

Vinyl tiles need to be acclimated to room temperature the same way as sheet vinyl. Take the tile out of the boxes and mix tiles from different boxes to give the floor a more consistent look. Whether they are being applied using adhesive or have a self-stick back, acclimation and mixing are necessary. Remove all transitions and moldings around the room.

INSTALLATION OF VINYL SHEET FLOORING

Resilient vinyl flooring can be installed using either adhesive or double-sided tape. As with any floor installation, clean the sub-floor thoroughly before applying adhesive or double-sided tape.

fastfacts

When the manufacturer packages the roll at the mill, it tends to stretch the wearlayer. This stretching is called stress, which has to be removed. This is done by back-rolling the amount being used prior to installation. Do this at the same temperature as the room it is being installed in.

DOUBLE-SIDED TAPE INSTALLATION OF ROLLED SHEET VINYL

Installation of rolled sheet vinyl can be done over any type of sub-floors previously listed—plywood, concrete, ceramic tile, and existing vinyl tile. Double-side tape installation is not recommended for areas covering in excess of 36 square yards or in rooms requiring more than one seam. Measure the floor and prepare the sheet for cutting. If the sheet has not already been rough cut by the supplier, take the roll out to a large, flat area. Driveways or garage floors are good areas to lay out the floor, once those areas have been cleaned of any dirt and stains which can get on the vinyl. Rough cut the sheet using a utility knife, leaving excess for cutting and wall irregularities. An over-age of 6 inches is usually sufficient. Now, bring the vinyl back into the room, and dry lay until the pattern is in the right place (Figure 6.5).

If you are using two pieces which must be seamed together, make adjustments for that match when the second piece is added. Continue to make cuts until the floor fits snugly, but not too tight, into all the corners. To make cuts on the inside and outside corners, take the utility knife and make relief cuts from the top down to where the floor and wall meet.

Once the vinyl has been cut down to the appropriate depth and fits snug in all the corners, it is time to trim the excess away. Take a utility knife and a carpenter's square and gradually trim any material lapped up on the wall. Do not cut away more than ⅛ inch clearance between the edge of the sheet vinyl and the wall (Figure 6.6).

Leave a gap between cabinets, pipes and walls of at least ³⁄₁₆ of an inch for expansion and seasonal movements. These gaps will be covered by the perimeter molding at the end of the installation.

Cut away vent and pipe openings after the perimeter of the room has been trimmed (Figure 6.7). To trim around pipes, make a paper template of the exact shape. Position the template around the shape, leaving several inches in all directions. Place some glue on the template and unroll the floor over the exact place it will be. Roll the flooring back carefully with the template stuck to it. Cut the shape out of the vinyl floor with a sharp utility knife.

Door casings can create a tricky cut, since there is no molding or grout to hide miscuts. Take your time, cutting the vinyl a little at a time, until it fits perfectly. To insure a secure fit, apply a thin bead of non-staining silicone around the door casing.

The other method is to do what is done for ceramic, carpet or wood flooring. Take a hand saw and cut the door casing to the thickness of the floor and tuck vinyl underneath (Figure 6.8).

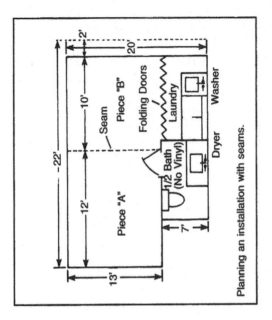

Planning an installation with seams.

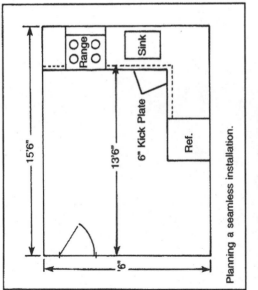

Planning a seamless installation.

FIGURE 6.5 Sheet placements and layout.

FIGURE 6.6 Trimming along perimeter walls.

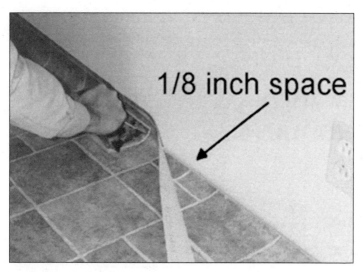

FIGURE 6.7 Cut out floor openings.

FIGURE 6.8 Cutting the door casing for floor placement.

Now that the floor has been completely laid and all the cuts have been made, it is time to secure it with the glas-tac acrylic double-sided tape. Do not tape around the entire perimeter of the room. It is only necessary to tape where there will be heavy movement or transition areas. Lift the floor at the transition area and place a strip of double-sided tape down (Figure 6.9).

Replace the vinyl down on top of the strip and roll over it with the hand roller. Place more tape where there will be seams and heavy furniture or appliances which will be moved at one time or another.

fastfacts

Accidental cuts can occur but there are ways to make them less noticeable. Cut a piece of no-stain double-sided tape and place it on the underside of the vinyl floor where the cut was made by mistake. Remove the back side of the tape and press down the vinyl sheet. Run a thin bead of seaming adhesive along the entire length of the cut on top of the vinyl, bonding the edges together.

FIGURE 6.9 Double-faced tape installation.

When a second piece is needed to complete the floor, the placement is done the same as the first piece. Check the design so it will match perfectly after all is laid in place and cut. Overlap the two pieces of vinyl after matching them up and cut through both at the same time (Figure 6.10).

Place double-sided tape down under the seam. Carefully replace both pieces of vinyl over the exposed tape, pressing down firmly on the seam. Take the hand roller and go over the seam line (Figure 6.11).

After it is secured down, seam sealer must be applied to the seam line (Figure 6.12). This sealer will activate the wearlayer of both pieces and create a chemical bond, making it one piece. Follow the manufacturer's directions exactly for application and safety for best results.

After the floor is cut and firmly in place and any seam sealer has dried, replace all moldings and T or E transitions at any doorways. If using quarter-round, raise it slightly above the flooring so it does not induce pressure on the vinyl.

When replacing the trims and transitions, nail them into the baseboard and not the floor. This will keep any pressure off the new floor and allow for natural expansion movements (Figure 6.13). Wait at least an hour before moving any appliances or furniture back onto new floor.

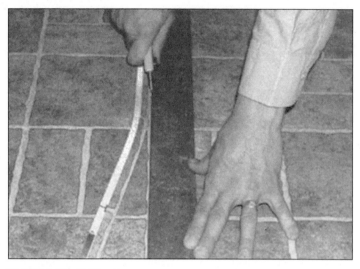

FIGURE 6.10 Cutting the seam.

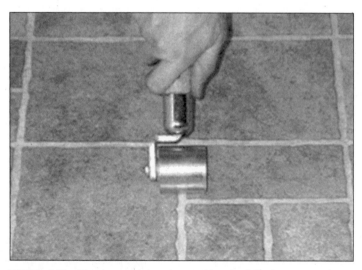

FIGURE 6.11 Place material over tape and hand roll firmly.

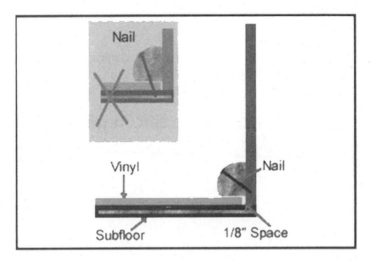

FIGURE 6.12 Apply sealer to seam of vinyl floor.

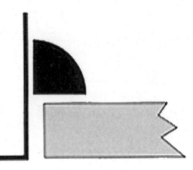

FIGURE 6.13 Trim replacement into baseboard.

INSTALLATION OF SHEET VINYL WITH ADHESIVE

Another option for installing sheet vinyl is with adhesive. This method is advised for single layer vinyl or heavy traffic areas such as a back porch. It is measured and cut the same way as the double-sided tape method, but will more strongly adhere to the sub-floor. The adhesive method will not allow for any expansion or seasonal movement once it is installed.

fastfacts

In extreme indoor conditions, like a prolonged period of cold weather, a sub-floor can dry out and shrink. This can cause the vinyl sheet flooring to buckle at the corners. If this occurs, remove the baseboard molding near the affected area, pull back the vinyl and re-trim. If the flooring shows a buckle or lump at a doorway, remove the transition and gently lift the vinyl from the tape to push out the fullness. Replace with a new piece of double-sided tape and press down firmly. Put back trims and transitions when you have a snug fit.

Measure and cut the sheet vinyl the same way as before and place in the room. Spread the vinyl adhesive on the sub-floor where the sheet will be placed. Do not place it directly on the back of the vinyl flooring.

Once the adhesive is spread with a smooth-edge trowel, roll the floor over it and press down firmly once the pattern is placed where you want it. Look for bulges where the floor has not properly adhered and flatten them immediately. By pressing firmly down with a roller, it should spread out any excess adhesive to the area that was missed and make a solid bond. Once the adhesive is dry and the floor is set, it is next to impossible to get the bumps out.

When a seam is required, follow these simple steps. Adhere the first piece of flooring down with adhesive, leaving a 10 inch space before the seam. Cut and adhere the second piece with at least a 2 inch overlay, or more if you need to make a pattern match. Spread adhesive over the rest of the floor, stopping 2 inches from the edge of the first sheet. Cut through both sheets where the seam will be. Lift up both halves and apply adhesive to the sub-floor. Press both halves in place and run a line of seam sealer along the top. Roll the entire floor out with a heavy roller, starting at the center of the room and working your way out towards the edges. Use a solvent to clean up any excess adhesive that has spilled on the surface of the vinyl, and replace any moldings and transitions around the room.

VINYL TILE INSTALLATION

Vinyl tiles can be placed over the same sub-floors as resilient sheet vinyl. Tile can also be placed over existing floors if a 15 pound felt paper is laid down first. Vinyl tiles can have cushioned or non-cushioned backing and may or may not have peel-back adhesive.

TILE AND SUB-FLOOR PREPARATION

Remove any dirt and stains from the existing floor, since adhesive does not stick to oil, grease or dirt. Whether the tiles are installed over concrete, plywood or wood sub-floors, make sure all areas are filled in and level.

Once the floor is clean and smooth, it is ready to be measured and centered. The tiles should be started at the center of the room. To find the center, measure the room and mark the center on each wall. If the room is not perfectly square, ignore the bump-outs and niches and mark accordingly. Take the chalk line and mark a line from the center marks on the walls. The point where the two lines intersect will be the center point (Figure 6.14).

INSTALLATION OF VINYL TILES

Once the center of the room has been determined, it is time to dry lay the tiles. Make sure the tiles have been acclimated to room temperature for at least 48 hours. Mix the boxes of tiles to give the whole floor a more consistent appearance. Dry lay the first row of tiles along one chalk line, then along the second chalk line (Figure 6.15).

fastfacts

To check for accuracy, measure 3 feet in one direction and 4 feet in the other direction at a 90 degree angle, and the distance between the 3 and 4 foot mark should be 5 feet. If it is not, measure again and mark your lines.

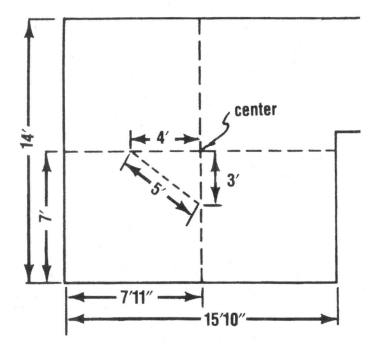

FIGURE 6.14 Location of center spot in a room.

Measure the distance between the wall and tiles. If the distance is more than 8 inches or less than 2 inches, re-mark and adjust your center mark. Move the center line that is parallel to the wall a half a tile closer, 4½ inches, so the wide row is less and the narrow row is more, making them close to even size rows. This will give the floor a more even look, rather than a very narrow row on one side and a wide row on the other (Figure 6.16).

Once the floor has been carefully planned and tile patterns are matched up, you are ready to apply the tile cement. If the tiles are peel-back adhesive, simply remove the paper backing and stick them to the floor. Press firmly and roll over them to make sure they are secured to the sub-floor. No tile cement is needed for this type of floor installation.

For non-stick tiles, mix a batch of tile cement according to manufacturer's instructions. Spread the cement over one-fourth of the floor with a notched trowel held at a 45 degree angle or a roller.

FIGURE 6.15 Dry lay the first and second row of tiles.

Start at one of the chalk lines and work within one quadrant at a time (Figure 6.17).

The cement should dry to proper consistency in about 15 minutes. Drying time will vary depending on the temperature and humidity of the room. If the cement is still sticky when you touch it, it is not quite ready. The cement should be tacky but not sticky, and then it is ready for installation.

Start with the first tile placed exactly at intersection of chalk lines. If the first tile is set incorrectly, it will throw the entire floor off. Butt each tile up against each other, leaving no gaps. Do not slide tiles into place, they must be laid on top of the cement and placed firmly into position. Lay the tiles alternately, toward each wall. This method will help counteract expansion and contraction of tiles while enhancing appearance. When you reach the end of the row and have to cut the tiles, lay a tile squarely over the last space to be covered. Mark a line and cut along the mark with scissors (Figure 6.18).

FIGURE 6.16 Re-mark the center line for more consistent end rows.

FIGURE 6.17 Spread cement tile one quadrant at a time.

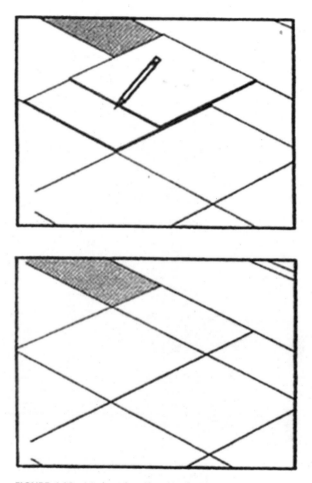

FIGURE 6.18 Mark and cut border tiles.

When obstacles such as pipes and vents need to be cut out, make a paper template of the area. Trace the pattern onto the tile and cut out the shape with a pair of scissors. Place the cut tile around the obstruction, making sure it is secured with enough tile cement (Figure 6.19).

When all the tiles have been cut and secured, roll the floor from the center out to the walls. If tiles are uneven, you may need to smooth them out by removing the tile and taking out excess cement

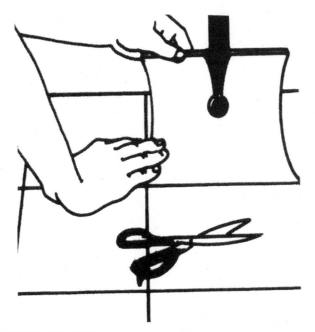

FIGURE 6.19 Cutting around obstructions.

underneath. Apply another thin coat of cement and replace the tile(s). If there is any excess cement on the surface of the tile, remove with solvent and let dry.

Even though there is literally a seam around each tile, there is no need to use seam sealer on any of them. The tile floor is made to look like squares and should be kept that way. Butting the tiles up against one another should result in a tight, consistent look.

Resilient vinyl tile is a fast and inexpensive way to get a new floor as it offers a variety of looks and durability. Following these instructions and tricks-of-the-trade will result in a beautiful, long-lasting floor.

BAMBOO FLOOR
INSTALLATION

Bamboo is a relatively new alternative in hardwood based flooring. This type of material blends a high quality floor with exceptional wear capabilities. The application of bamboo lends itself to areas where there is extreme wear and tear. Rooms suitable for bamboo flooring include kitchens, bathrooms, and family rooms. The natural resistance to water lends itself to rooms where moisture can occur on the surface.

The natural appearance of bamboo sets it apart from the sometime harsh grain of hardwood floors. It offers a warm ambiance in the winter and coolness in the summer and is comfortable to walk on. Cost wise, bamboo is a reasonable choice when compared to other hardwood floors. Bamboo is a much harder product than some hardwoods, making it an even better choice economically. Bamboo is as hard as hard maple and 50 percent more stable than red oak.

Compared to wood, bamboo has a higher density of fiber making it resistant to wear. Most bamboo floors have a UV anti-scratch top coating making it almost impossible to scratch.

Bamboo flooring is a more stable product than hardwood. It offers less expansion and contraction than any common hardwood flooring, including oak and maple. When it comes to installation, bamboo planks are similar to hardwood floors. They can be installed by glue-down, nailed or floating method.

Most bamboo planks have several layers of polyurethane applied with an ultraviolet hardening process. So clean-up is easy, simply using a damp mop or a mild wood cleaner will help maintain this floor for years to come.

THE BACKGROUND ON BAMBOO

Bamboo is an extremely fast-growing plant; some can grow over three feet per day. There are one thousand species of bamboo, all with root systems that help it re-grow after it has been harvested for many decades over. The Moso species is the type of bamboo used in most flooring planks. It is the hardest of all the species, harvested after four to six years of growth. When harvested it has usually reached a height of 90 feet and a diameter of 8 to 12 inches.

Bamboo is a fast growing, renewable resource, making it a good choice ecologically. The species harvested is not used for feeding by animals; therefore it does not deplete a life-sustaining food source for wildlife.

A flat strip is milled from the core of the wall of bamboo. These strips are then boiled in boric acid and a lime solution. These solutions extract the starch that attracts termites and powder post beetles. This non-toxic repellent make the strips pest free and remains that way for the life of the product.

The strips are then kiln dried and sanded to a smooth finish and then laminated edge to edge to create a single-ply panel. These panels are then laminated again to each other creating multi-layer bamboo plywood strips.

TYPES OF BAMBOO FLOORS

There are two types of bamboo floors available, horizontal or vertical, both in plank lengths only. Both are the traditional tongue and groove construction with matching ends. Standard planks are 3⅜ inches wide, with thickness of ¹⁵⁄₃₂ to ⅝ inch and lengths of 36 or 72 inches. This flooring has a honey finish with a natural mottled pattern inherent to bamboo.

Horizontal Bamboo

Horizontal bamboo or flat-pressed flooring has the appearance of wide lines running through it (Figure 7.1). It has a rather open look and tends to give any area a roomier look.

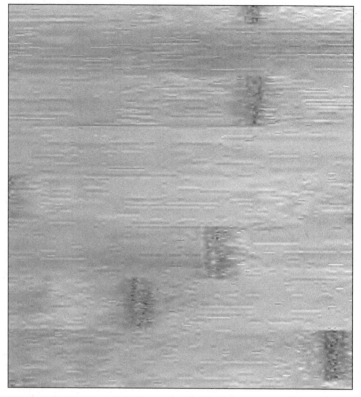

FIGURE 7.1 Horizontal bamboo construction.

Vertical Bamboo

Vertical bamboo or side-pressed flooring has a more dense appearance than the horizontal (Figure 7.2). This tends to give the floor a more consistent appearance.

COLOR OPTIONS FOR BAMBOO FLOORS

Bamboo, being the natural product it is, has only two options for color. They are natural (light) and carbonized (dark) as shown in Figures 7.3 and 7.4. However, bamboo can be ordered unfinished

FIGURE 7.2 Vertical bamboo construction.

and have any color applied to it using a water-based paint or stain. Once the color is dry, several coats of polyurethane should be applied. This will still have the hardness and durability of natural or carbonized bamboo.

MEASURING FOR BAMBOO FLOORING

The measurement process for bamboo is similar to hardwood flooring. Figure out the room's square footage by multiplying the length by the width. If installing a floor in rooms with odd shaped areas, divide those areas into sections and calculate the area of each. Add all the dimensions together to come up with the total square footage

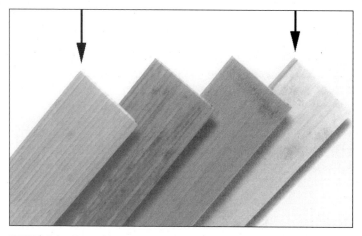

FIGURE 7.3 Natural horizontal and vertical bamboo.

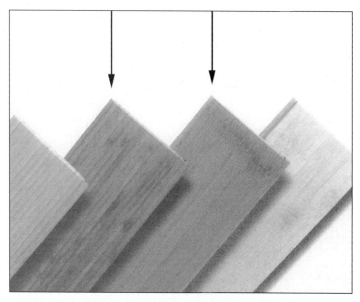

FIGURE 7.4 Carbonized horizontal and vertical bamboo.

needed to cover the entire area. Take the total number of inches and divide by 144 to find square footage. Be sure to add ten percent for waste and errors. See Figure 7.5 for a simple chart on how to calculate room area.

Bamboo planks are sold in bundles which cover a specific amount of square footage. Read the label to find out how much area each box will cover.

TOOLS NEEDED FOR BAMBOO FLOOR INSTALLATION

Here is a checklist of tools needed for the installation of bamboo flooring. This list includes tools for measuring, cutting and installation.

Tool Checklist

✔ Hardwood floor nailer

✔ Rubber mallet

✔ Tenon saw, circular saw or hand saw

✔ Electric drill with a $\frac{3}{32}$ inch (2mm) bit

✔ Claw hammer

✔ Nail punch

✔ Square

✔ Measuring tape

✔ Pry bar

✔ Five inch putty knife

✔ Chalk line

✔ Flooring screws

✔ Construction paper or 15 pound felt floor liner

✔ Hardwood flooring nails (1¼ or 1½ inch)

✔ Screw shank finishing nails (2 inch) or regular finishing nails

Not all of these items will be used for every installation. Depending on the sub-floor used and the method chosen for installation, some tools will not be necessary. Make sure all of these tools are handy so no delay is caused during work on the project.

Square Footage

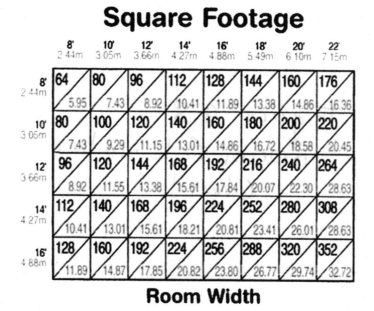

	8' 2.44m	10' 3.05m	12' 3.66m	14' 4.27m	16' 4.88m	18' 5.49m	20' 6.10m	22' 7.15m
8' 2.44m	64 5.95	80 7.43	96 8.92	112 10.41	128 11.89	144 13.38	160 14.86	176 16.36
10' 3.05m	80 7.43	100 9.29	120 11.15	140 13.01	160 14.86	180 16.72	200 18.58	220 20.45
12' 3.66m	96 8.92	120 11.55	144 13.38	168 15.61	192 17.84	216 20.07	240 22.30	264 28.63
14' 4.27m	112 10.41	140 13.01	168 15.61	196 18.21	224 20.81	252 23.41	280 26.01	308 28.63
16' 4.88m	128 11.89	160 14.87	192 17.85	224 20.82	256 23.80	288 26.77	320 29.74	352 32.72

Room Width

FIGURE 7.5 Room calculation chart.

fastfacts

Bamboo flooring is a naturally colored product. It is not stained, so there will be definite color variations. Natural colors will range from a white to a more yellow tone. This may happen within one carton or from carton to carton. If you chose a carbonized finished, the bamboo planks are steamed to give them an even darker appearance. There is nothing wrong with the planks' finish, it is a natural finish and not a defect that the customer must be made aware of before installation. Work from several cartons at once for best results.

SUB-FLOORS

Bamboo flooring can be installed over standard sub-floors consisting of plywood or concrete. Specific care should be taken when installing over either product. It is not recommended that bamboo be installed over floors with radiant heat. This will cause the flooring to expand and contract due to temperature changes.

PLYWOOD SUB-FLOORS

Plywood sub-floors are a great surface for bamboo floors to be installed over. Be sure the plywood is at least ⅝ inch or thicker and that the corners are staggered. Nail down all the way around the perimeter of each piece of plywood. When nailing is completed, use the nail punch to bring nails flush with plywood surface. If there are any gaps or holes in the plywood, fill them in with wood filler and let dry. Check for levelness, making sure no bumps or ridges exist. Sand down ridges or add leveler to any low areas in the floor. Also, remember to leave an expansion gap around the perimeter of the room. This gap should be no more than ¼ inch. Sweep the floor until it is free of dust and dirt.

Check for moisture, making sure moisture content does not exceed 12 percent for proper installation of bamboo floors. If the moisture is too high, delay the installation and turn up the heat or increase ventilation in the room.

CONCRETE SUB-FLOORS

Concrete sub-floors can also be used for bamboo installation. The important factor to keep in mind before installing over concrete is excessive moisture. Concrete floors, especially below-grade, can have high moisture content. This can be taken care of by first testing the floor, making sure it has moisture content of below 12 percent. If moisture is a problem, install rolls of 15 pound felt floor liner over entire area. Then install a plywood sub-floor over felt liner. If moisture is under control, you may glue bamboo flooring directly to concrete. As with any sub-floor check for levelness, filling in any holes and grinding down any ridges or high areas before installing bamboo.

PREPARATION BEFORE INSTALLING BAMBOO FLOOR PLANKS

The humidity level in the room is one of the most important factors in bamboo floors. Bamboo is a living material that will change with the humidity levels. Humidity levels must be maintained between 40 and 50 percent year round.

If installing in new construction, the heating system must be operational for at least a week at 71 degrees. All plastering and concrete work must be dry. Bamboo will absorb moisture from anything that has not dried in a room. Any type of moisture in a room will also affect the humidity level making it too unstable to install the bamboo.

Bamboo flooring should be taken out of the boxes and placed in the room to acclimatize for at least 24 hours before installation. Mix the planks from several cartons together to give the floor a more natural look.

Remove all the baseboards if no sub-floor had to be added first. Remove and save any trim and transitions. Take a handsaw and cut away the bottom of any door frames ¾ inch to make room for planks (Figure 7.6).

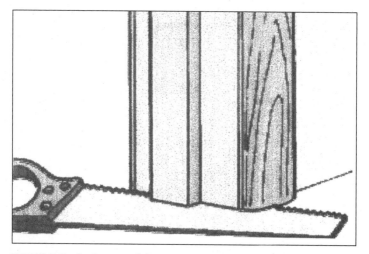

FIGURE 7.6   Cut bottom of door casing.

INSTALLATION OF THE BAMBOO FLOOR

Whether the floor is installed using the nail-down method or glued down, you must determine which direction the bamboo planks will lay. Install the strips parallel to the longest length wall in the room. Try to install them in the opposite direction of the floor joists. If that's not possible, lay them at a 45 degree angle.

NAIL DOWN INSTALLATION

Start with a chalk line coming out 3 inches from the longest wall for planks measuring 2¼ inches wide (Figure 7.7). This should leave room for the tongue of the plank and expansion space around the perimeter of the room.

A chalk line should be started 4 inches out from the wall for strips measuring 3¼ inches. This chalk line must be at a 90 degree angle to the adjacent wall. This first guide line must be exact since it will set the line for the entire project.

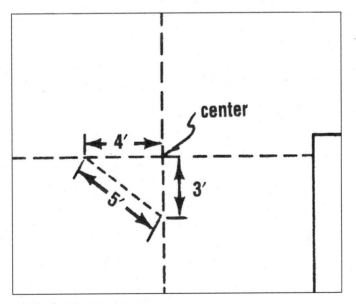

FIGURE 7.7 Mark a chalk line.

Select the strips for the first three rows and dry lay them out on the floor in the general pattern in which they'll be installed. Choose the straightest strips for these rows. Take your time when selecting each strip to make the floor look natural and well planned. If there are any flaws, cut them and make any adjustments. If some strips are noticeably darker than others, place them in areas where they will be less noticed.

For the first row along the wall, lay the tongue edge of the strip along the guideline, leaving room for the expansion gap along perimeter of wall (Figure 7.8). The baseboard or molding will cover any gaps when the project is completed.

The first row must be secured to the floor using screw shank flooring nails (or finishing nails). Drill holes 1 inch from the edge, and at 12 to 16 inches apart into the strips. Nail into these holes to avoid splitting the bamboo strips. The first few rows must be hand nailed rather than with the hardwood floor nailer because of the wall obstruction. When there is enough clearance, use the power nailer with 1¼ or 1½ nails.

Measure and cut the strips to complete the first row. Any remaining sections from the first row should be used to start the second row to minimize waste. The strip selected for finishing the first row should be long enough to leave enough to start second row. Leave ¼ inch between the wall and the end of each strip in every row (Figure 7.9).

Start the second row with a strip at least 6 inches shorter or longer than the strip used in the first row. This will avoid any joints from lining up or clustering. Set the strips in place, and drill a hole on the tongue edge at a 45 degree angle every 8 to 10 inches apart. Using screw shank flooring nails, secure the plank in place (Figure 7.10).

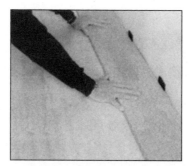

FIGURE 7.8 Lay the first strip tongue
side along guide line.

LEEDS COLLEGE OF BUILDING
LIBRARY

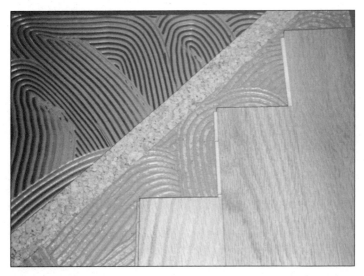

FIGURE 7.9 Leave a space of ¼ inch between wall and end of each plank in every row.

FIGURE 7.10 Secure planks every 8 to 10 inches with screw shank nails.

The subsequent rows must be installed in the same manner. Once the wall has been cleared, use the power nail gun to speed up the process. Go ahead and lay out six to eight rows of boards to mix shades and lengths. This will ensure that the end joints won't meet and the appearance of the floor will look natural. Complete nailing the working strip flooring first before nailing the following row. As

fastfacts

When using a manual hardwood floor nailer, there are some basic tips that can make the job a little easier. Set the nailer down on a piece of plywood or cardboard to keep it from damaging the new floor. Also, if you maintain a standing position while using the nailer it will give you more strength when driving the nails in. Nails which do not go completely in can be driven in by a hammer and a nail punch. Be sure the nailer base is placed square on the edge of the strip to avoid damaging it.

you move along, take cardboard and spread across the finished floor so as not to damage it while installing the rest of the floor.

When it comes to obstructions in the floor, such as vents and pipes, simply make a template from paper and attach it to the back side of the strip that will be placed in that location. Using a handsaw, carefully cut the pattern out and lay the plank in place (Figure 7.11). This plank is installed the same way as any of the other planks.

When you reach the last rows, hand-nailing must be used to install the remaining planks because of the wall obstruction. The drill, screw shank flooring nails, and the nail punch will help secure the planks. For the last row, you may have to cut the plank if the wall is not square. Take the plank and lay it over the area it is going into and make a mark along the length. This will give you the angle needed to cut and fit that last row in place. Use a circular or hand saw to make the cut and secure in place.

Once the floor is installed, replace all baseboards, moldings, and vent covers. Vacuum the floor thoroughly and lightly spray bamboo floor cleaner on a terry cloth towel. This will help remove any loose dirt or soil. Do not spray directly on the bamboo floor. It is also a good idea to save a few planks for any repairs which may occur due to damage or climate changes in the future.

GLUE DOWN METHOD

The glue down method involves the same care and cutting as the nail down method. An expansion gap around the perimeter of the

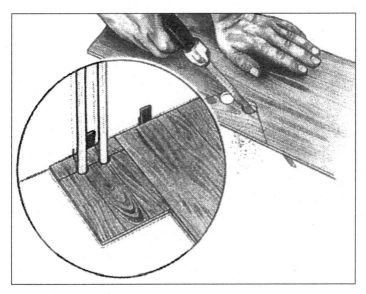

FIGURE 7.11 Cutouts for obstacles.

room is still necessary to allow for expansion. Gaps between the wall and the end of each strip in each row must be maintained at ¼ inch. Start the first row along the longest wall, making a chalk line mark away from the wall according to the width of the planks.

2¼ inch plank width = 3 inch gap from wall around perimeter.
3¼ inch plank width = 4 inch gap from wall around perimeter.

You may glue down bamboo flooring over concrete or plywood sub-floors using a urethane adhesive. Be sure to follow the manufacturer's instructions on the container. Use the recommended trowel size and dry time. Only apply adhesive in area you are working on. Do not get ahead of the bamboo and let the adhesive dry before you install flooring over it. If you apply adhesive 6 to 8 rows ahead, this should allow plenty of time to measure, cut, and install before it dries. If adhesive is drying too fast, apply it in a smaller area.

If you are installing bamboo flooring directly onto concrete, make sure it has been tested for moisture content. Recommended tests for this are:

fastfacts

A bamboo floor is not unlike a hardwood floor. It is affected by humidity and temperature, so great care must be taken before installing in a room that has a width of more than 20 feet. Rather than starting on the longest wall, start at the center of the room. Nail down the starter strip and work your way out from both directions, adding additional expansion spaces as needed. This method allows the floor more room to expand and will help avoid warping and cupping later on.

- *Calcium chloride test*
- *Polyfilm test*
- *Phenolphthalein test*
- *Pin or pinless moisture meter*

Finish the last row of bamboo planks by making proper square cuts if necessary. Replace any baseboard and trim, along with transitions.

SANDING AND REFINISHING BAMBOO FLOORS

Bamboo floors can be sanded and stained like most hardwood floors. Use the same techniques and sanding patterns with drum, belt and edge sanders. Water or oil-modified finishes will work well on any sanded bamboo floor. Any type of polyurethane finish can be applied once a paint or stain has dried. Always follow the manufacturer's product instructions on the can for best results.

Protect the floor against direct sunlight. Remember it is a natural product and it will be affected by light. Over time any intense light, either natural or artificial, may discolor bamboo flooring. The lighter the floor, the more noticeable the discoloration will be. You can protect the floor by using rugs, rearranging furniture over different areas of the floor, and using drapes which block out UV sunlight.

Never use wax, oil-based detergents, or other household cleaners on bamboo floors. These may cause a glassy finish and make the floor slippery and difficult to clean. Use a very mild detergent or a lightly damp mop for cleaning and maintaining the floor.

Bamboo floors are easy to refinish or change the color after a few years. Like any floor, bamboo lasts longer when it is maintained and protected.

Bamboo floors have not been available for a long time, but seem to have no bad reviews. As long as you and the customer are aware of the natural variety of coloration, this type of flooring is a great alternative to hardwood flooring.

8

BRICK FLOORS

Most people don't think of brick as a true source for floors. Over the last decade, however, brick has become a great source for designers and architects who want to offer a different and more substantial look for their clients. Brick is becoming more common in residential projects rather than strictly contract applications. Along with being an obvious choice for walls, the floor has become the next level for brick.

While you would use a complete brick for most walls, a new product which uses only the top or side portion of the brick is being applied to floors. This option lets you use brick without worrying about the load on the sub-floor. Similar to ceramic tile, the weight of the brick must be determined before installing over any sub-floor.

Brick floors are a good idea in areas where there is a lot of traffic, such as foyers, kitchens and laundry rooms. Once properly sealed, brick holds up well under water, wear-and-tear, and decades of use.

Cost-wise, brick is more expensive than most floor applications, however, it is will last a lifetime when maintained properly. Take into consideration one floor for a lifetime or replacing the same floor several times. Maintenance after installation is basic, simply sweep and damp mop for quick clean-up.

TYPES OF BRICK

Generally you can use any type of standard brick for a flooring project. The key is to use only the top ½ inch of any brick for the actual
installation. If the brick is too porous when cut to a thin layer, it may
become too brittle to use on a floor. So choose a brick with a
smooth surface.

Brick can be baked clay or terra-cotta, fired in an oven at 1500
degrees Fahrenheit. Depending on the finish, brick is mainly a
porous material and will need to be sealed for protection from
stains. If it is a terra-cotta brick, use a terra-cotta sealer and use a
non-yellowing polyurethane sealer for a high gloss finish.

The colors of brick are pretty traditional; generally you will find
16 main finishes of color used in the industry. Depending on your
project, keep in mind this floor will last a lifetime so choose a color
which will transcend fads and trends (Figure 8.1).

The size of each brick tile is the size of a standard brick in length
and width. The length is 7⅝ inches, the width is 2¼ inches and the
depth is ½ inch. You can also use the brick tiles for baseboard trim
and steps coming up to a floor. Look at the chart to see the options
available (Figure 8.2).

ROOM MEASUREMENT FOR BRICK TILE

Measure the area in which the floor is to be installed and find out
the total square footage. Take the length of the room and multiply
it by the width. This measurement will help in determining how
many bricks will be needed for the installation. Remember to add
an additional five to ten percent for cutting and mistakes. Depending on the pattern and layout of the brick work, four tiles will generally cover one square foot. This includes room for a ⅝ inch grout
line between the bricks.

TOOLS NEEDED FOR BRICK TILE INSTALLATION

Here is a list of tools needed for the installation of brick tile. When it
comes to grout and adhesive products, be sure to follow the manufacturer's instructions on each container for best results.

Tool Checklist

✔ Metal square

✔ Chalk line

✔ Tape measure

✔ One-quarter inch notched trowel

✔ Wet tile saw with diamond blade

✔ Rubber mallet

✔ Grout float

✔ Level

✔ Thin set mortar

✔ Short-nap roller

✔ Brick sealer

✔ Grout

✔ Plastic brick spacers

All of these tools will be necessary to complete this project. When it comes to sub-floors, additional tools and supplies may be needed.

PREPARING THE SUB-FLOOR

Brick floors can be installed over concrete or plywood sub-floors with the proper preparation. Keep in mind that the brick floor, when complete, will be extremely heavy on the sub-floor, so strong supports are necessary.

Concrete Sub-floors

If you are installing brick tiles over a concrete floor it must clean and level. It is important to check for high moisture levels before installing any flooring over concrete. A moisture barrier may need to be installed over the concrete before the tile adhesive is applied (Figure 8.3). This may also apply to concrete sub-floors which will have plywood sub-floors placed over them.

Once the concrete has been leveled and moisture-proofed, if necessary, you may apply adhesive directly and begin to lay the brick (Figure 8.4).

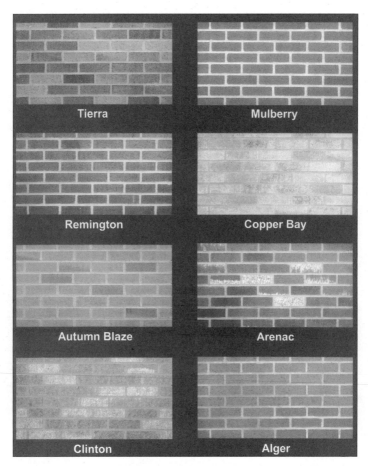

FIGURE 8.1 Brick color choices.

Installation over existing or new plywood is another option for sub-floors. Installation of plywood is the same as with any other type of sub-floor. Use AC grade plywood, making sure the corners are staggered and that the nails are set and filled in to achieve a smooth and level surface.

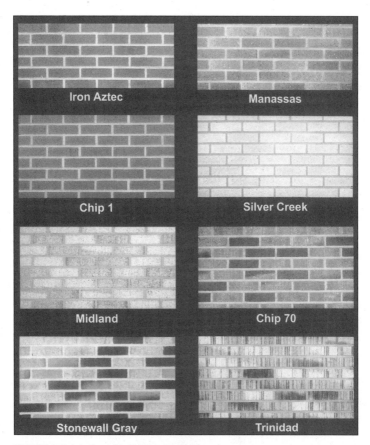

FIGURE 8.1 *(continued)* Brick color choices.

FLOOR LAYOUT

Brick can be laid out in a variety of patterns, including a running bond (Figure 8.5), a basket-weave pattern (Figure 8.6), a ladder-weave (Figure 8.7) and a herringbone pattern (Figure 8.8).

While any pattern will give a room a custom look, some are more complex than others. Be sure to choose a pattern which you can realistically install and which works with the style of the room.

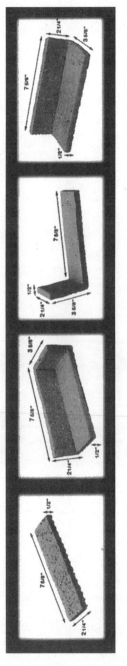

FIGURE 8.2 Brick tile options.

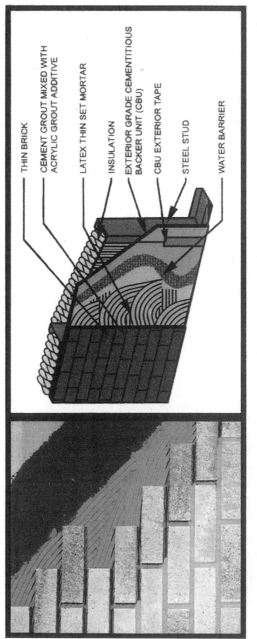

THIN BRICK

CEMENT GROUT MIXED WITH ACRYLIC GROUT ADDITIVE

LATEX THIN SET MORTAR

INSULATION

EXTERIOR GRADE CEMENTITIOUS BACKER UNIT (CBU)

CBU EXTERIOR TAPE

STEEL STUD

WATER BARRIER

FIGURE 8.3 Application of moisture barrier over concrete floor.

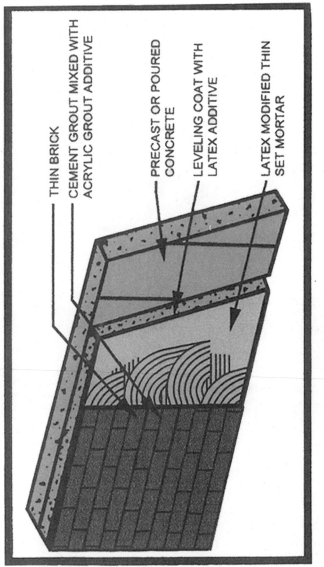

THIN BRICK

CEMENT GROUT MIXED WITH ACRYLIC GROUT ADDITIVE

PRECAST OR POURED CONCRETE

LEVELING COAT WITH LATEX ADDITIVE

LATEX MODIFIED THIN SET MORTAR

FIGURE 8.4 Direct application of brick on concrete sub-floor.

FIGURE 8.5 Running bond pattern.

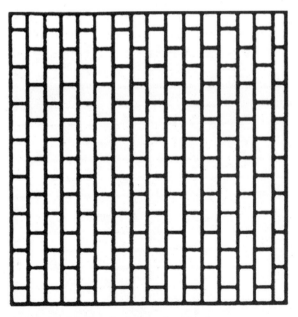

FIGURE 8.5 *(continued)* Running bond pattern.

FIGURE 8.6 Basket-weave pattern.

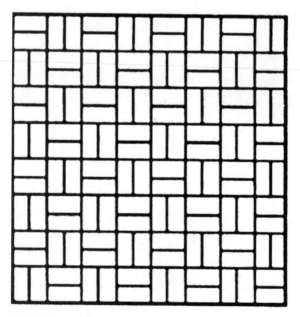

FIGURE 8.6 *(continued)* Basket-weave pattern.

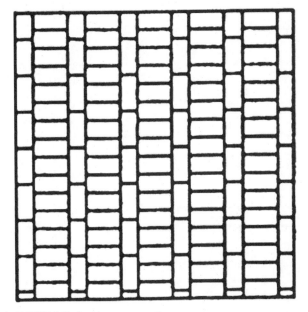

FIGURE 8.7 Ladder-weave pattern.

FIGURE 8.8 Herringbone brick pattern.

FIGURE 8.8 *(continued)* Herringbone brick pattern.

BRICK FLOOR INSTALLATION

Locate the center point of the room by measuring the longest wall of the room, marking the center. Then measure the room's width, marking the center. Run a chalk line across both center marks making that intersection of lines the center point.

Make sure at this time you have decided upon a pattern for the floor. Dry lay a few pieces to get the idea of how it will go down before mixing and applying any adhesive onto the sub-floor.

Mix a batch of thin-set mortar to the consistency of toothpaste. There is no need to let it set-up. Once you have the consistency, begin applying to the sub-floor. Don't mix more than you can install in 20 to 30 minutes of time. Mortar can be very dusty so consider doing this outside or in a garage that is well ventilated. A latex-modified mortar is recommended for a stronger bond to the brick, it also give you a little more time to work with the mortar.

Spread out the mortar onto the sub-floor with the ¼ inch notched trowel. Do not apply the mortar anywhere on the brick tile themselves. Hold the trowel at a 45 degree angle to produce a level bed of mortar with grooves to help hold the bricks in place (Figure 8.9).

Begin laying the bricks along any guideline starting at the center of the room. Press them firmly into place to secure them to the mortar bed. Follow the chalk line exactly. If the first bricks are even slightly off, the entire floor will be out of alignment. Be sure to leave a ⅜ inch of space between each brick for the grout. Use the plastic spacers to insure an exact space (Figure 8.10).

When the brick has to be cut, use the wet saw with a diamond-impregnated blade, which is cooled by water and helps provide a very smooth edge on each cut.

Cutouts are cut with a wet saw according to the shape they will fit around. Take your time when cutting and measuring since brick is a very volatile product which can chip or crack if you go too fast or cut to close to an edge. Make allowances for all of the spacing and layouts before installing.

When the floor has dried, remove any spacers and apply a coat of brick sealer with a short-nap roller. This will help prevent the bricks from absorbing too much moisture from the grout. After applying the sealer, allow the floor to dry for at least 24 hours before applying grout.

FIGURE 8.9   Spread mortar with ¼ inch notched trowel at a 45 degree angle.

FIGURE 8.10 Use plastic spacers for a ⅜ inch grout line.

Once the sealer has dried, apply the grout much the same way as you would for a ceramic tile floor. You can choose from two types of grout for the brick floor. Standard sanded tile grout looks very neat and is easy to work with. This is what is used for most ceramic tile floors. The other option if you want it to look like a real brick floor is to use actual brick mortar. If you go with the brick mortar, use the type S mortar and sand with three shovels of sand to one shovel of mortar with water. Only mix enough to work 30 minutes at a time. Use the back side of the trowel to spread the grout into the joints in all different directions. After applying a section of the grout, let it dry for a few minutes then wipe of excess with a clean, damp sponge. Make the joints smooth with the sponge, rinsing it and changing water frequently. When all the grout has been applied, wipe off any haze with a damp sponge.

Before a grout sealer is applied, it is recommended that the floor be cleaned with muriatic acid. This will make sure that any areas the grout has not covered will be protected from moisture and that any excess grout on the brick will be washed off before it is sealed in for life.

fastfacts

When cutting out corners with the wet saw, raise the brick as you push it toward the center cut for a precise square edge. It is also a good idea to soak the bricks before cutting them. This will cut down on most of the dust bricks throw off.

When it is time to seal the brick, choose from a water-based sealer or a non-yellowing polyurethane. If a water-based sealant is used, you can apply this immediately to the floor. If you use a poly-urethane sealer, you must wait 30 days for the brick to dry. These products are highly flammable, so be sure to work in a well venti-lated area and to dispose of all materials properly when finished. Test a small area of the brick before applying the sealer to the entire floor. Certain sealers will darken the brick after application. Apply with a paint brush around the edges first, then roll on the remain-ing sealer with a short-nap roller. Always follow the manufacturer's instructions on the container for best results.

Brick floors are an incredible alternative, offering a great variety of designs and finishes. Installation can be time-consuming but well worth the effort when the project is complete.

9

LEATHER FLOORS

We have been dealing with the more traditional types of flooring throughout this book, but now there is a unique flooring product which is making quite a statement lately. Leather floors are becoming more commonplace in everyday residences. Leather used to be saved for the more upscale homes and offices; even some luxury yachts are covered in leather from floor to ceiling. No longer reserved for furniture, leather possesses excellent qualities for flooring.

Leather flooring can be installed in just about every room of the home. It complements any room which has wood tones running through it, as well as contemporary settings, giving it a great contrast. There are, however, a few exceptions for placement of leather floors, for example, a kitchen where there is a possibility of dropping hot pans or food. Another example, are areas where there is a possibility of tracking in a lot of salt from outside through foyers or laundry rooms. Salt would permanently alter the finish of the hide.

This type of floor is a natural when it comes to sound absorption. It makes a great floor for music rooms, home theaters and libraries. Unlike the type of leather used in clothing, these tiles are hard to the touch. Through a process of tanning and dying, these tiles are able to withstand the wear and tear of years in use.

Cost wise, leather floors are comparable to hardwood floors or custom carpet in cost and it is installed much like ceramic tile. It is a unique choice which is not for everyone or every look, but it is beautiful nonetheless.

Maintenance for leather floors is basically nonexistent. It is not suggested that you clean the floor at all, only an occasional waxing for keeping up the patina. This floor will darken with time and any stain or irregular mark will only enhance the appearance.

THE LOWDOWN ON LEATHER

The leather used for flooring has been put through a special process of tanning and dying. The leather used for tiles is from the center of the hide, otherwise known as the "bend" (Figure 9.1). Aniline dyes penetrate all the way through the bends when the hide is placed in wooden drums filled with color.

FIGURE 9.1 Bend of the hide.

This is a time-consuming process which ends up protecting the leather tiles inside and out. If the floor was ever cut or scratched, the same color on the surface would be seen all the way through the tile. No tile is perfect, making it the most natural looking floor available.

SIZES AND SHAPES OF LEATHER TILES

There are a variety of sizes available for leather tiles. Generally tiles are available in seven different shapes and sizes (Figure 9.2).

- *18 x 18 inches square*
- *12 x 12 inches square*
- *8 x 8 inches square*
- *4 x 4 inches square*
- *8 x 4 inches*
- *2 x 8 inches*
- *2¾ x 8 inches*

Patterns vary for leather floors, but generally there are nine styles which are commonly used in conjunction with the seven standard tile sizes and shapes (Figure 9.3).

- *Piazza*
- *Trompe l'oeil*
- *Herringbone*
- *Squares*
- *Parquet*
- *Pavimento*
- *Harlequin*
- *Castello*
- *Chevrons*

There are hundreds more custom patterns which can be used by using the standard tile shapes. Cutting and spacing tiles according to the shape of a room can give that room a whole new look.

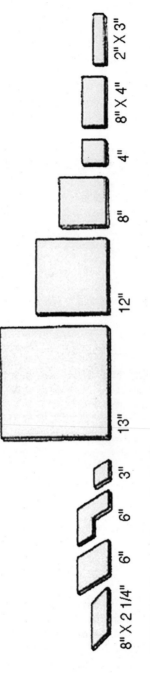

FIGURE 9.2 Sizes and shapes of leather tile.

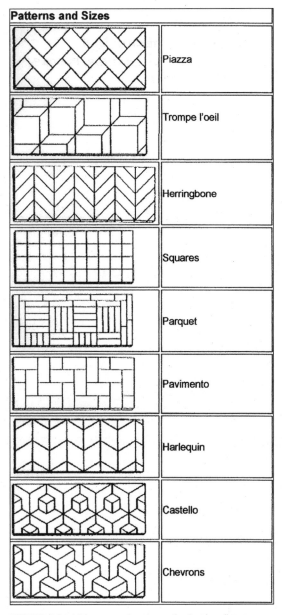

FIGURE 9.3 Nine general patterns for leather floors.

SUB-FLOOR PREPARATION

Leather tiles can be installed over any type of sub-floor as long as it properly prepared. The best type of sub-floor to use is plywood. Whether installing over concrete or plywood, all seams and holes must be clean, level and completely flush.

Leather is a somewhat thin product and any little deviation in the sub-floor will show. If there is any problem with moisture coming through the sub-floor it is not recommended for leather installation. Even if you were to place a moisture barrier under the sub-floor, it is not a good idea, since even one leakage through the floor would ruin the leather. It is far costlier to replace the floor than not to install over any area in danger of moisture. Never apply tiles to a finished surface, as the adhesive will apply to the finish rather than the actual sub-floor.

MEASUREMENT OF AREA FOR LEATHER TILE

Determine the total square footage of the area to be installed. Of course, it all depends on the size and shapes of tiles you will be using. After determining the square footage, lay out your pattern and see how many tiles it takes to cover a one foot square area. Multiply that by the total square feet to come up with the total amount of tiles needed to complete the project. Allow an additional 5 to 10 percent for waste and mistakes.

TOOLS FOR INSTALLATION OF LEATHER TILE

What follows is a checklist of tools and supplies needed for leather tile installation.

Tool Checklist

- ✔ Tape measure
- ✔ Chalk line
- ✔ Square
- ✔ Utility knife
- ✔ Non-Flammable contact adhesive

✔ Brush to apply adhesive

✔ Carnuba wax

All of these tools and supplies will be necessary for the installation of these tiles. Additional supplies and tools may be necessary if a new sub-floor has to be installed.

INSTALLATION OF LEATHER TILES

Before installing the tile, open all the boxes of leather and let them acclimate to the temperature of the room for at least 48 hours. Find the center point of the room by measuring along the longest wall and marking the center point. Next, go to the wall beside the longest wall and find the center point. Take the chalk line and make a line out from the center point of both walls. The intersection of the two lines will be the center point. Always start the tile pattern at the center of the room and work along the chalk lines (Figure 9.4).

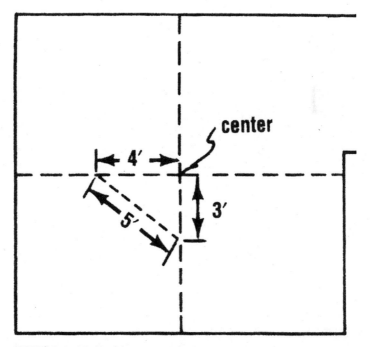

FIGURE 9.4 Marking the center point in a room.

Dry lay a few squares to get the idea of how to place them correctly. Use a high quality contact adhesive and apply it directly to the sub-floor. Do not apply the adhesive directly to the leather tiles. Be sure to read and follow the manufacturer's instructions on the adhesive for best results. Never apply more adhesive than the area you can work in. Contact adhesive dries quickly, so work in an area about 2 x 3 feet until you get the hang of it. Leather tiles are not perfect so don't try and force them to fit. Natural is the look that you are trying to achieve. Be sure to leave a minimal gap between each tile of no more than ⅜ inch. Leather tiles do not expand or contract, so a gap around the perimeter of the room is not necessary.

When there is a tile or an obstruction to be trimmed, simply make a template of the cutout and place it over the leather. Then, cut it with a utility knife and secure it to the floor with the contact adhesive.

When the floor has been completely installed, go over the entire surface with carnuba wax. This wax acts as the grout, filling in any gaps between the tiles (Figure 9.5). Buff the floor a minimum of three times before using it. This will ensure a rich and durable finish. Depending on use, buffing every three weeks and waxing several times a year is all that is needed.

FIGURE 9.5 Apply carnuba wax to leather tile floor.

Installation of leather tiles takes time to master. If you consider all the sizes and shapes available, you may never have the opportunity to install all the options available. This is a fairly easy but expensive installation, so take great care when cutting and have fun with the patterns. Leather floors make a great change from the traditional floors that are available today.

LEEDS COLLEGE OF BUILDING
LIBRARY

10

CLEANING, MAINTENANCE, AND REPAIR

After the work of installing the flooring is completed, the most important information for the home owner is understanding how to clean and maintain the new floor. If the original look and finish is to last, cleaning and maintenance should be the highest priority.

In this chapter we will discuss everyday cleaning and some common maintenance and repair tips. What might seem like simple tasks can really make the difference in the life of a floor. Using this information will enhance the appreciation of your work, and ensure a long floor life.

CARPET MAINTENANCE

Most carpet is manufactured as cut-pile or loop-pile. The style and quality of a carpet makes a difference in regards to cleaning and maintenance. Cut-pile is made from tightly wound fibers which are dyed only on the outside and have a white or cream core. This type of carpet will show wear early on if not maintained as vigorously as a loop-pile. Loop-pile is dyed throughout the entire fiber, which makes it look better and last longer.

While almost all carpet is treated with some sort of stain resistant solution, it still requires regular cleaning and care. Those stain resistant solutions eventually wear away and will either have to be reapplied, or the carpet will have to be replaced much sooner than necessary.

EVERYDAY CARPET MAINTENANCE

Let's start with what can be done to protect the carpet from wear and tear, and dirt buildup. The best thing you can do for carpeting is to vacuum it at least once every week. Carpet is well known for being a great dirt trapper, so when vacuuming, be sure to adjust the vacuum to the proper setting for the depth of the pile. Use the edging tool of the vacuum to work along all the edges of the carpet, not just the center. If the carpet is not cleaned along the edges, eventually those edges will become very dirty and be extremely difficult to clean. Remember to move furniture once in a while and clean under it as well.

When it comes to high traffic areas, such as doorways or traffic paths through a room, area rugs can be placed over the carpet to keep the dirt to a minimum. Don't forget to vacuum the area rugs when you vacuum the carpet.

CARPET CLEANING

Carpet cleaning is recommended about every six months, although this will vary depending on traffic and use. Two common methods that can be used are wet cleaning and dry cleaning. Most carpet can be cleaned either way, but there are good and bad issues with both. Dry cleaning is a less messy way to clean and is usually recommended for carpet that does not require heavy cleaning.

The dry method involves the use of detergent foam sold in boxes which cover a specific amount of square feet. This method is less harsh on the fibers because it goes on fairly dry and vacuums up within an hour of application. If you just want to lightly clean and renew the pile on the carpet, this is probably the best method.

Wet cleaning is a good method for deep cleaning a carpet. You can do this job yourself or hire a professional cleaning service. Wet cleaning can be done using the water extraction method, which requires a cleaning machine that injects a cleaning solution into the carpet, and then extracts the water and dirt like a large shop vacuum. The other wet method applies a cleaning solution, and then utilizes at least two rotary brushes to scrub and lift the dirt.

The downside of wet cleaning is that you can get the carpet too wet and cause damage to the padding underneath. Go over the carpet just once with the wet machine, spraying pre-spot on areas that are extremely dirty.

When using pre-spot cleaner, be sure to try it on a less conspicuous area first in case it discolors the carpet. If you use a brush for applying the spot cleaner, be careful not to overdo it so as not to fray or wear out the fibers. Always use a soft bristle brush and rub gently.

COMMON CARPET REPAIRS

Chances are good that, over time, carpet will be subjected to some sort of damage or stain due to normal everyday use. Here are some of the common repair and maintenance methods available.

Carpet Burns

Most burns in a carpet are surface related and can be simply trimmed off with a small scissors. In the case of severe burns from cigarettes or fireplace tools, the carpet may need to be patched. Tools needed are a sharp utility knife, hot-glue gun, and a carpet tractor.

Use the utility knife to cut out the damaged section of carpet, making the cuts as straight as possible, and cutting only through the backing and not the fibers. Make the cut out at least two inches away from the burn mark. Take the cut out patch and use it as a template to cut the repair patch from a spare carpet remnant. If no remnants are available, small patches can be made from carpeting from inconspicous areas, such as a closet corner.

Use the hot-glue gun to set the patch in place, press it down carefully, and then glue around the edges so the fibers won't pull out. Use the carpet tractor to roll the seam and blend the fibers so the seam doesn't show.

This method can also be used when it comes to any type of stain that cannot be taken out with standard cleaning, and can also be employed in areas where the carpet is torn.

Wrinkled Carpet

Wrinkled carpet is another common problem that can happen for a number of different reasons. If carpet is not stretched properly, carpet can come loose along the edges of the tack strips. Another possible cause is excessive moisture in the carpet that can cause the carpet to expand, shrink, and expand again unevenly. Some carpeting of low quality will simply stretch and wrinkle over a period of time.

The best way to get rid of the wrinkle is to have the carpet completely re-stretched. If you pull the carpet up and back only to the point where the wrinkle is to re-stretch it, the carpet in the entire room will be affected. In a short time the wrinkle will probably return unless the whole room is completely re-stretched.

CARPET THRESHOLDS

Areas between rooms usually have a transition strip of metal or wood placed between them to keep the carpet from coming loose. There are different types of transition strips for areas going from carpet to tile or carpet to wood. Over time these transitions can become worn or loose and will need to be replaced. These strips are readily available through carpet or building suppliers.

Most transition strips come in standard lengths to fit specific dimensions, or in long strips which can be cut to fit any size transition. If applying a transition strip between carpet and tile or carpet and cement, I suggest laying a thin layer of caulk under the strip and then attaching the strip with screws. This is an inexpensive and quick way to make your carpet look fresh and new.

Seam Repair

Seams of a carpet may start to come apart after excessive use, and occasionally with natural wear, depending on the quality of carpet. You can take up the carpet where the seam is loose, and remove the old seam tape that was originally used to hold the carpet together. Seal the edges of the carpet with seaming glue to prevent the carpet threads from coming loose. Cut a new piece of seaming tape to the right size, centering it under the carpet seam. Use a carpet seaming iron which has been adjusted to the appropriate setting and place it directly on the seam tape. Allow it to melt the glue on the tape for 30 seconds. Slowly move the iron along the seam tape, pressing the seam in place as you go. Let the glue set for at least 24 hours before walking on it.

CARPET STAIN REMOVAL

Before we discuss stain removal, it is wise to bear in mind that stain removal can do more damage than good. Some commercial products designed to remove stains can actually bleach or remove color from the carpet. Many store brand cleaners contain bleach additives that can bleach the dye right out of the carpet.

Carpet dying is usually reserved for a highly experienced carpet dyer. If you try to use common dyes, such as fabric dye, hair dye, or shoe dye, chances are good that the color will not precisely match the carpet color, and the repair will probably fade quickly. Below is a list of stains and some remedies for removing them from carpet.

STAIN	REMEDY
Rust	A white vinegar or light detergent solution
Lipstick	Nail polish remover, dry cleaning fluid, detergent solution, ammonia or white vinegar solution
Pet Stains	Detergent solution, white vinegar or ammonia solution
Coffee	Warm white vinegar solution or detergent solution.
Blood	Use a cold spot setting solution to avoid setting the stain. Then use an ammonia solution with a warm water rinse.
Candle Wax	Place a clean white towel over dried candle wax, run an iron set on low over towel. Repeat using a clean area of the towel until all the wax is gone.
Beverage	Detergent solution, ammonia solution or white vinegar
Nail Polish	Nail polish remover or detergent solution
Crayon	Dry-cleaning fluid or detergent solution
Gum	Soften the gum with hair dryer then lift it off the carpet with a plastic glove. Use dry-cleaning fluid for any remaining residue.
Chocolate	Dry-cleaning fluid, detergent, ammonia or white vinegar

Ink	Nail polish remover, dry-cleaning solution, detergent or white vinegar solution
Latex Paint	Detergent or ammonia solution
Shoe Polish	Nail polish remover, dry-cleaning fluid or detergent solution

A properly maintained carpet should last for at least 10 to 12 years. Quality and use are major factors in the life of a carpet, and cleaning and maintainance are just as important. Sometimes carpet can be damaged so badly, as in the case of fire and water damage, that no amount of repair or cleaning can bring it back to life.

If carpet is exposed to flooding and becomes totally saturated, there is a way to salvage it. One thing to remember: once carpet has become totally saturated with water, the padding underneath must be thrown out and replaced. Padding cannot be dried out without resulting in the production of mold and bacteria. It is well worth the time and effort to replace padding immediately. Attempting to dry and salvage soaked carpet padding is inevitably a waste of time.

Water can be removed from carpet by two different methods. The first method is to use a shop vacuum or rent a carpet cleaner designed to extract water from the rug. Regular vacuums should never be used for this task . They are not designed for such heavy duty use, and create an electrical hazard with exposure to excessive moisture.

If a wet vacuum is used, you should use it to remove all of the water possible. If the water has gotten into the padding, the vacuum will probably not be able to pull the water through to the carpet surface. If the water has reached into the padding, remove the carpet by raising the corners of the carpet one at a time. Take the carpet completely out and let it dry. Then remove the soaked padding and dispose of it. Make sure the floor underneath the carpet pad is allowed to dry thoroughly. If the surface is still wet, wipe it up right away. A wet floor can quickly buckle up and become loose and uneven. Then replace the soaked padding with new padding.

If the carpet is wet only on top, slits can be cut into non-conspicuous areas and the hose attachment of the vacuum can be slipped into the slits to blow air underneath the carpet. This will cause the carpet to float, and create an avenue for venting the moisture.

It is very important to get the dirty water out of the carpet before it dries. If the carpet is allowed to dry without vacuuming out as much moisture as possible, mold and mildew will be almost impossible to remove.

RESILIENT VINYL FLOOR MAINTENANCE

The care and maintenance of a vinyl floor will make a difference in the length of time it will last. Vinyl floors are quite sturdy when it comes to wear and tear, but they can look old and worn before their time if they are not properly maintained. In this section I'll show you the simple steps to take for everyday care. For vinyl floors that have seen better days, I'll tell you about some great solutions that can make the floor look new again.

EVERYDAY VINYL FLOOR MAINTENANCE

The first step to taking care of any vinyl floor is to keep it swept with a soft bristle broom. Don't use a straw broom. Straw brooms have harsh ends that can scratch or dull the finish of the floor. Straw brooms are also not very consistent when it comes to picking up dirt or grit completely. A good soft bristle broom with a straight edge is the best tool to use.

If the floor is in a heavy traffic area, which is usually the case with vinyl floors, dirt and grit is often regularly tracked in, and can get ground into the top layer of the floor. If the floor is swept every day, the grit can be kept to a minimum and less mopping and waxing will have to be done.

Besides sweeping, another great preventative measure is to take your shoes off before you walk across the floor. If a vinyl floor is located off a garage or black top driveway, the bottoms of shoes will pick up small bits of the black top and other debris and embed into the floor. Over time this will create a yellow path across the vinyl and will be impossible to remove.

Area rugs or runners are another preventative method to keep the stains and dirt from getting through the surface. Do not use rubber-backed rugs. Rubber backed rugs can discolor vinyl over time.

The next step in vinyl floor maintenance is mopping. Depending on the floor surface, there are various commercial solutions available to remove lightly ground in dirt and surface stains.

A no-wax floor is much easier to clean and remove stains from, but be sure the cleaner used is compatible with the floor. All cleaners are not compatible with all floors, and some can cause irreparable harm to the finishes. Be sure to avoid abrasive detergents or cleaners that might scratch the floor. Some cleaners also leave scratchy abrasive residue behind after they are wiped up. These abrasive particles can be spread across the floor by the mop or sponge and actually cause damage to the floor. A simple sponge, slightly dampened with a mild solution of ammonia and water, are all that is necessary to give a vinyl floor a good cleaning.

Remember to move furniture or appliances out of the way to sweep and mop under them. If left uncleaned, those areas will eventually fade, making it difficult to rearrange appliances or furniture to another part of the room without exposing differences in tile color.

Another issue with vinyl floors is the dent and scratch factor. Since vinyl is a soft product, it can tend to dent when appliances or furniture is placed on top of it. When moving appliances in or around the room, place rugs or plywood under them and carefully walk the appliances to the new position. Use floor protectors under each leg to avoid dents and scratches. The heavier the piece, the wider the leg protector should be (Figure 10.1).

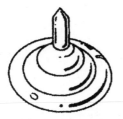

FIGURE 10.1 Floor protection leg options.

Stain Removal

Following is a list of stain removal remedies for resilient vinyl floors.

STAIN	REMEDY
Heel Marks	Non-abrasive household cleaner, rubbing alcohol. Light scuff marks can be erased with a pencil eraser
Paint	If still wet, use mineral spirits. If dry, scrape with a thin, plastic spatula
Magic Marker	Mineral spirits, nail polish remover, or rubbing alcohol
Candle Wax	When dry, scrape with a thin plastic spatula
Tar	Citrus-based cleaner or mineral spirits
Shoe Polish	Citrus-based cleaner or mineral spirits
Crayon	Mineral spirits or commercial cleaner
Grape Juice	Full strength bleach
Pen Ink	Citrus-based cleaner, mineral spirits, or rubbing alcohol
Rust	Oxalic acid and water, one part acid to 10 parts water. (Follow manufacturer's instructions)

After using any of these treatments, be sure to rinse the area with a damp sponge to remove all traces of residue.

COMMON VINYL FLOOR REPAIRS

Common vinyl floor damage can range from holes made by the legs of appliances and furniture, dents from appliances, and wear from excessive traffic. Curling or tearing of the vinyl from everyday use can also occur.

A hole or dent in a vinyl floor can be repaired easily by carefully lifting the vinyl and applying clear silicone underneath. This fills in the dent or hole, and makes matching the vinyl a non-issue.

If a seam of the vinyl has come loose, a seam repair kit can be purchased from a floor supplier. Lift the seam up and clean it thoroughly, then mix the two-part sealer according to the manufacturer's instructions. Apply it along the edge of the seam. Place a weight, such as a telephone book, on the seam so it adheres properly until it is dry, which will usually take 24 hours.

If an entire tile is damaged, it can be replaced by matching a spare tile and cutting it according to size and pattern. Remember that if the floor has been down for a long time, it may be difficult to find tile that will match the existing tile. It is always a good idea to set aside a few extra tiles after a floor job is completed.

Take a piece of new tile and lay it over the damaged area. Trim out a larger piece of new tile than the damaged area, and tape it down over the damaged piece. Make sure that the pattern matches, and then cut through both pieces with a utility knife. Remove the newly cut piece and then remove the damaged tile. This can be done by warming up the damaged piece with a hair dryer or carefully prying it up with a thin scraper.

Be sure not to damage other surrounding tiles. Once the old tile is removed, apply adhesive to the floor and carefully put the new piece of tile in place. Tape and weigh it down until it has set and is completely dry. Give the repair 24 hours to set before exposing the area to traffic.

Curling vinyl, which can occur in areas of excessive moisture, such as often occurs around the edges of bathtubs, can usually be simply repaired by using a hair dryer to warm up the vinyl until it is softened enough to be flattened out. Apply a thin line of flooring adhesive and press the vinyl back in place. Place a weight on it until it is dry.

Vinyl tile will hold up well if cared for properly, but sometimes it needs extra maintenance to prevent the flooring from giving out before its time. Follow the simple maintenance plan and you should enjoy the floor for many years.

LAMINATE FLOOR MAINTENANCE

Laminate floors are designed to be extremely durable and hold up under excessive wear, but they must be properly maintained to look their best. Let's cover some basic tips for everyday care and then list some stain remedies that can be used.

Routine cleaning for a laminate floor includes vacuuming, dusting with a cloth, and occasional wiping with a damp mop. There are some great damp mopping products on the market today that make this job quick and easy. Do not get the floor too wet, because excessive water may penetrate the edges and grooves of the floor, and eventually create unnecessary problems.

Do not use soap-based detergents to clean a laminate floor, or "mop and shine" products. These products will only dull the floor and make it tough to get the shiny finish that the floor originally had. Do not wax or polish a laminate floor, as this can make the floor extremely slippery and dangerous to walk on.

Do not use abrasive cleaners or tools to clean the floor. Cleaning tools and products such as steel wool or scouring powder will scratch the floor.

If damp mopping with water is just not enough to clean the floor, use a mixture of one cup vinegar to one gallon warm water. An alternative cleaning solution is ½ cup household ammonia to one gallon warm water.

PREVENTATIVE MAINTENANCE

Some simple things can be done to prevent damage and extra cleaning to a laminate floor. Place colorfast rugs over high traffic areas to protect the floor from abrasive particles like asphalt or sand.

Using felt floor protectors on the bottom of legs of furniture prevents possible scratches from occurring. If any furniture has metal casters, consider replacing them with rubber wheels. Also, avoid dragging any furniture over the floor. Always lift or move the furniture using a rug or felt pads.

Pets can create a problem on laminate floors. Keep their nails trimmed regularly to minimize scratching. Also, clean up any accidents as soon as they occur.

It is a good idea to remove your shoes before walking on a laminate floor to avoid scuff marks and dents. Shoes attract and carry a lot of dirt and abrasive material that can be tracked over a floor. High heel shoes are especially harmful to laminate floors. Wipe up any spills as soon as they occur. The longer the spill sits on the floor, the greater the possibility of staining.

LAMINATE FLOOR STAIN REMOVAL

Following is a list of common laminate floor stains and their remedies. After removing any stain, rinse the area to remove any residue.

STAIN	REMEDY
Chewing Gum	Let harden and scrape with a blunt plastic scraper
Candle Wax	Scrape up carefully with a plastic spatula

Lipstick	Paint thinner or acetone
Nail Polish	Acetone-based nail polish remover
Tar	Acetone or denatured alcohol
Grape Juice/wine	Lukewarm water and ammonia
Paint or Varnish	Wipe with mineral spirits or water while still wet. If dry, scrape up with a thin plastic spatula.
Pen Ink	Acetone or paint thinner
Crayon	Rub out with a dry cloth or acetone if necessary
Chocolate	Warm water and non-abrasive cleaner
Cigarette Burns	Acetone, nail-polish remover, or denatured alcohol
Rubber Heel Marks	Rub with a dry cloth or acetone if necessary

COMMON LAMINATE FLOOR REPAIRS

If a scratch occurs in a laminate floor, it can be filled in with a wax crayon designed for just this purpose. The color of the floor can be matched choosing from a variety of crayon colors. If the scratch is too deep for a crayon to cover it, you can also use floor putty. As with wax crayons, floor putty is also available in a variety of colors. Simply clean the scratch and apply enough putty to fill in the damage. Most laminate floor manufacturers sell both the crayons and floor putty that match their product.

If the damage is larger than ½ inch, replacement of the entire plank is recommended. Since laminate floor planks do not fade, color matching should not be an issue.

CERAMIC TILE MAINTENANCE

Ceramic tile is perhaps the easiest flooring to care for. Along with the tile, there is grout cleaning and maintenance to think about. Ceramic tile is resistant to stains, but grout can absorb and harbor stains and mildew.

Daily or weekly maintenance of tile floors should include sweeping, vacuuming, or using a non-oily dust mop. Damp mop as needed, using a pH neutral tile cleaner. Do not use abrasive cleaners or tools to clean the tile, such as steel wool or metal tools. A good solution to use for cleaning ceramic tile is one gallon warm water, one tbsp. borax and two tbsp. clear ammonia.

Ceramic tile suppliers can recommend the right cleaners and brushes to use for the specific tile involved. Always follow the manufacturers' instructions for any cleaner being used. Once tile has been rinsed, allow it to dry before walking or working on it. If a tile is sealed properly, the steps listed above are all you will need to do.

If you need to reseal tile, remember that it must be stripped before a sealer is re-applied. Choose the stripper recommended by the sealer manufacturer and follow all instructions carefully.

PREVENTATIVE MAINTENANCE

Be sure to use the right cleaning products that clean the tile without causing staining. Ask the tile manufacturer which product is best for the tile that was installed. Always clean up any spills right away so they do not penetrate or become ground into the tile or grout.

Once the tile has been cleaned and rinsed, make sure there is no excess water that can get into porous grout or tiles . Over time, excess water can cause tiles to loosen.

GROUT CARE AND MAINTENANCE

Grout care is basically the same as tile care. Grout is usually sealed so it can resist most stains, but grout also periodically requires deeper cleaning. If the grout is exposed to excessive moisture, the moisture can be absorbed and become difficult to remove.

Non-abrasive tools should be used, such as a grout brush or toothbrush, to apply grout cleaner. There are correct cleaners for every grout.

With any cleaners, be sure to test on an inconspicuous area first. Some tile cleaners may have an adverse affect on other surfaces in the room, such as wood or metal. Read directions before using any general cleaner.

CERAMIC TILE STAIN REMOVAL

Following is a list of remedies for some common tile stains.

STAIN	REMEDY
Rust	Use a commercial rust remover made for fabric, then use a household cleaner to remove any additional residue
Nail Polish	Nail polish remover
Gum	Harden gum by holding ice on it, then remove with a thin plastic spatula
Grease	Use a commercial spot lifter
Cigarette Burns	Gently scour with steel wool. After burn is removed, polish the tile with paste wax.
Oil-based Products	Use a mild solvent such as mineral spirits, then household cleaners mixed with poultice
Ink, Blood, Fruit Juice	Three percent hydrogen peroxide or a non-bleach cleaner
Coffee	Non-bleach cleaner

Poultice, which is a mixture of plaster of Paris and hydrogen peroxide, can be a great cleaning solution for pulling out stains from very porous stone tiles.

Another method for removing stubborn stains from the surface of grout is sandpaper. Fold a piece of sandpaper and gently rub it back and forth along the grout line. If that doesn't do the trick, try a pencil eraser.

GROUT REPAIR

Sometimes simply cleaning the grout cannot get rid of stubborn stains. In extreme cases, the grout may have to be removed. There is a specific tool, called a grout saw, which can be used to dig out the original damaged grout. Once this task is complete, replace grout with the proper color using a grout float. Wipe off any excess with a damp sponge and allow to dry.

When the grout is completely dry, apply a grout sealer which will help resist future stains and keep the repair easier to clean.

CERAMIC TILE REPAIR

It is rare that a tile has to be replaced, but if necessary, it is easy enough to do. If the tile is chipped or has been cracked, it is better to take it out than fill it in or try to cover it up. If the tile is not completely cracked, score the tile first and then break it up with a hammer. Be extremely cautious of the surrounding tiles so as not to damage them. Carefully remove the broken pieces of the tile and clean off any excess adhesive from floor.

Apply new adhesive to the floor and tile. Tap the new tile into place and let it dry. When the adhesive has dried for 24 hours, apply grout to the clean joint. Be sure to apply sealer to both the tile and grout to finish the repair.

Remember the issue with this repair is that, unless a tile was saved from the original job, the color of both the tile and new grout may not match the others. It is always a good idea to save some remaining tile from any tile job for this particular reason.

HARDWOOD FLOOR MAINTENANCE

Hardwood floors can last for up to 50 years if maintained properly. There are several things one can do to prevent wear and tear on a hardwood floor, so that you can achieve the best long term results.

In this section we will list simple care tips and some common repair methods for hardwood floors.

PREVENTATIVE MAINTENANCE

How a hardwood floor is cleaned should be determined by whether it has a surface finish or is waxed. A simple test to determine this is to drop a little water in an out-of-the-way place on the floor. If a white spot appears after about ten minutes, the floor is waxed. Remove the white spot by buffing with the finest grade of steel wool and a little wax. A basic surface finish floor maintenance list follows:

- *Dust or vacuum regularly.*
- *Clean up any spills as soon as they occur.*
- *Damp mop the floor only once or twice a year. Remember: water and wood don't mix well.*
- *Use a cleaning product recommended by the floor manufacturer.*

- *Do not wax the floor. Wax will give it a dull finish and make it very slippery.*
- *Use chair glides or felt pads under the legs.*

A basic waxed finish floor maintenance list follows:

- *Vacuum and dust regularly.*
- *Wipe up any spills as soon as they occur.*
- *Clean the floor once or twice a year with a solvent-based cleaner, then wax and buff the floor to protect it for another year.*

HARDWOOD FLOOR STAIN REMOVAL

A variety of stains can occur on hardwood floors. Following is a list of hardwood floor stains and remedies for repairing them:

STAIN	REMEDY
Dye	Use one part chlorine bleach and two parts water
Ink	Warm water and mild detergent
Rust	Commercial rust remover
Urine	Hot, damp cloth with mild scouring powder. Use a 10 to 1 solution of diluted liquid bleach.
Blood	Sponge up with clear, cold water. Use a solution of ammonia and cold water for more stains.
Grease/Oil	Remove as much as possible with newspaper, or paper towels, then place a cloth saturated with dry cleaning fluid on the stain for five minutes. Wipe the area dry and wash with detergent and water.

The first time you use any of the remedies, test it first on an inconspicuous part of the floor to make sure it will not damage the finish. Always read the labels of cleaners and solvents and follow directions. Be sure to ventilate the room properly when using any mixture of cleaners.

Scouring powders can help remove any stain, but may scratch the floor. If you feel you have to use such powders, be sure to use the mildest variety available. Concentrated liquid commercial household cleaners rubbed onto a stubborn stain may remove the stain as effectively as using scouring powder, without the danger of scratching. Rinse all areas off well, recoat with wax, and buff.

HARDWOOD FLOOR SURFACE REPAIRS

Surface problems can occur with hardwood flooring. Following is a list of problems and their remedies.

PROBLEM	SOLUTION
Seasonal Cracks	Increase the humidity in the dry season; install a humidifier
Dents	Cover with a dampened cloth and press with an electric iron
Scratches	Wax the area or use a thin coat of dusting spray rubbed into the scratch
Bubbles	Using two types of finishes may result in bubbles on the surface of the finish. For light damage, lightly sand and recoat. If heavy bubbling, sand, stain and refinish.
Cigarette Burns	Burnish with fine steel wool or scrape the burned area. Wax or sand, stain and refinish.
Water Spots	Buff lightly with the finest grade steel wool, then wax. If spot does not come out, use fine sandpaper, stain, and recoat.
Heel Marks	Wood cleaner or wax
Pet Stains	Wood cleaner followed by mild bleach or household vinegar
Ink	Wood cleaner followed by mild bleach or household vinegar
Mold, Mildew	Floor cleaner. If wood fibers are stained, remove and refinish

HARDWOOD FLOOR REPAIRS

If a hardwood floor has been flooded, there are a few quick things to do to stop or slow the action of the water from creating an even larger problem. Remove all excess water from the floor immediately. Fans and dehumidifiers will help speed up the process. If the floor is in a home with forced air heat, turn off any humidification and heat the residence to 76 to 80 degrees. Let it flow continuously until the floor has completely dried out. If no permanent staining has occurred, recoating of the floor with urethane may be all that is needed.

If there is some staining from water damage, re-finishing may be necessary before recoating the floor with urethane. If the water has been removed and cupping of the floor is still noticeable, lightly sand directly across the grain. Make sure not to sand all the way down the bare floor.

Nail down any loose areas and fill in any cracks. Apply a coat of stain according to manufacturer's directions and let dry. After the stain has properly dried, apply a urethane finish to protect the floor from future damage.

If these methods don't work, then the floor is probably too damaged from the water, and should be entirely replaced. If you try to nail down and sand areas which are severely damaged, they will only pop up again in the future and continue to be a problem.

Before deciding which method is right, let the floor settle for at least two weeks to see what damage has actually occurred. Just make sure you have taken off all the water as quickly as possible to allow it to dry thoroughly.

Cupping and crowning of wood floors are natural occurrences due to humidity and moisture in the house (Figure 10.2). Kiln-dried wood boards that are subjected to moisture only on one side will expand only on that side, warping and bending away from the moist side. Cupping occurs for one reason only, the gain or loss of moisture on one side of the board faster than the other. If the boards are dried quickly, they will return to the normal flat position. However, if they are cupped for a long period of time, the boards will remain that way. Cupping is a more common problem in wider boards.

Cracks are probably the most common problem in wood floors. Hairline cracks are considered normal in 2 ½ inch wide floor strips, if they close up during non-heating months and are not any thicker than the width of a dime. There are primarily three reasons for the cracks appearing in wood floor planks.

- *Foundation settlement. When outside walls, or the center supports under the house's center beam settle, the area of the floor actually stretches. This causes cracks over joints in the plywood sub-floors.*

FIGURE 10.2 Cupping of wood floors.

- *Over-drying above forced air heating ducts. If the cracks are limited to areas in the hallways or other areas above the heating plants, check for insulation irregularities.*
- *Improper sub-floor material. Nail holding capability is imperative to the floor installation process. If the sub-floor does not hold nails, cracks can occur from abnormal moisture absorption or heavy traffic.*

If the floor has a surface finish, matching wood filler should do the trick for repairing small cracks. Be sure to give the area a coat of polyurethane after the fill has been allowed to dry. If the floor has a wax finish, use the finest grade of steel wool, and clean and wax over the area that has been filled in. If the flooring is cupped or crowned, these methods will not be effective.

Panelizing is another problem which can occur with hardwood floor planks. Due to uneven distribution of moisture, random areas of planks will shrink and remain tight together. This appearance is very noticeable and requires the attention of a very experienced floor repair person.

Panelizing is not caused by the manufacturer, or whether the wood has or has not been kiln-dried. It can be caused by many external forces such as: tugging, pushing and locking strips together. What follows are some causes of panelizing of hardwood floors.

Foundation Settlement

Perimeter foundation settlement can cause a traditional joist floor structure to stretch across the center beam of a house, resulting in cracks in surface floors. Cracks or panelizing usually occur near the center beam and are often limited to one or two major cracks.

Floor Finish Edge-Bonding

Some types of polyurethane floor finishes seep through the flooring strips and literally glue the boards together. When the floor loses moisture, the floor boards can shrink and the joints can break apart at the weakest point

Sub-floor movement

Sub-floor material may be exposed to elements of the weather before they are installed. Sub-flooring can take in moisture before it is installed and then shrink, causing nails to loosen and resulting in a shift in the floor, making cracks appear. Sub-floor movement is probably the most common cause of panelizing.

Panelizing Remedies

It's a mistake to pull up and simply replace a panelized floor. The problem is not with the floor, but rather with the sub-floor.

A number of options are available to help remedy panelized floors. One is to remove or replace selected boards where large areas of panelizing have occurred. When possible, replace the boards in the same sequence as they were removed. This will allow for a simpler finish repair.

The other option is fill in the cracks of the panelized boards by creating filler from the sanding dust of the same or another floor. Mix this with a paste and the same stain pigment used to color the floor originally. Be sure to reseal the floor after any filler has been used, then wax if necessary.

If the cracks are ⅛ inch and wider, any filler will be squeezed out as the flooring expands in the humid season. Replacement of the boards is the best solution.

Another problem with wood floors that can be caused by refinishing is called chattering. This occurs when the sanding drum used to take off the surface finish runs unevenly over the floor. A series of round marks appear on the floor and are difficult to remove. This can be avoided by checking the the sanding machine for the following possible problems:

- *Check wheels for out-of-roundness, worn axles or embedded particles*
- *Check wheel bearings for roughness*
- *Check rear casters for excessive up and down play*
- *Check the condition of the drum*
- *Check drum shaft, motor shaft, and fan shaft for roughness and for excessive play*
- *Check pulley alignment, check belts for hard spots or unbalance*
- *Check for worn shafts or worn pulleys*
- *Make sure sandpaper has not been applied too tightly or loosely. Either can cause chatter marks.*

If all the checks fail to solve the chatter marks, replacement of the drum is probably necessary.

GENERAL FLOOR REPAIRS

Floor repairs are generally very simple to work out. In this section I will go into a few of the most common repairs. Whether it is a squeaky floor or replacing a floorboard or two, this section will show how it is done.

SQUEAKY FLOORS

Most homes eventually settle and force the floors to move with it. This can cause nails or screws to loosen and create a squeak. These squeaks can occur in a hallway, bedroom, or stairway. Common causes of squeaky floors can include:

- *Warped boards that rock when they are walked on*
- *Cheaply manufactured floor strips with ill-fitting tongue and groove*
- *A sub-floor separated from joists due to settling*
- *Weakened joists that have dried out, rotted, or have faulty bridging*

To resolve the squeak you must first locate the exact location of the noise. Once located, there are some simple ways to fix it.

Shim the sub-floor with wood shims between a joist and loose sub-floor boards. Gently tap the shims in place, making sure not to widen the gap you're trying to fill. After the noise has been elimi-nated, add some adhesive to the shim to help it stay in place.

Cleating the sub-floor is another method to get rid of squeaks. Place a cleat along the joist supporting the loose boards (Figure 10.3). Prop the cleat in place with a piece of 2 x 4 so it lies snugly

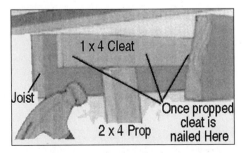

FIGURE 10.3   Place a cleat along the joists.

against the joist and sub-floor. Use an 8d nail to nail the cleat to the joist. Once the cleat is firmly in place, remove the 2 x 4.

Bridging the joist can get rid of squeaks over a large area. The joist beneath the floor may be shifting and unable to support the floor properly. Stabilize the sub-floor by attaching steel bridging between the joists (Figure 10.4). Use as many as needed to stabilize the floor and quiet the squeaks.

Installing screws from below the sub-floor is a very quick and easy way to quiet the squeaks. Drill a pilot hole the size of the screw shank. Insert a screw through a large-diameter washer, and turn it into the pilot hole (Figure 10.5). As the screw turns it will pull the floorboards down, tightening them.

FIGURE 10.4 Bridging joist.

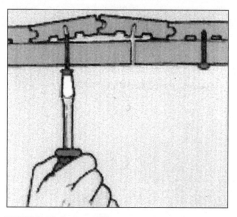

FIGURE 10.5 Installing screws from below sub-floor.

Nailing into the floor finish can be another way to force the floor to tighten up. If you are unable to get under the sub-floor, this is a great method to get through the floorboard and into the sub-floor from the top (Figure 10.6). You can nail right through carpet or hardwood as long as you are sure to do it in a discreet location.

Glazier points can also be placed between boards that are rubbing together. If lubricants do not work first, then rub some glazier points with graphite and set them between the boards, using a putty knife, about every six inches (Figure 10.7).

If you discover any loose boards anywhere, whether they squeak or not, they can be repaired the same way. Sometimes flooring products may have to be pulled back or removed to get to a loose board. Replace any flooring product in exactly the same way it was installed.

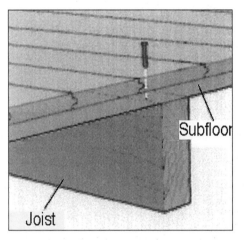

FIGURE 10.6 Nailing through the finish floor.

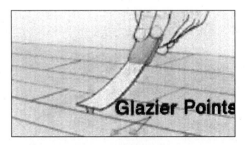

FIGURE 10.7 Setting glazier points.

Sometimes the quick fix just won't work, and an entire floor or floorboard must be completely removed. It is always a good idea to save any extra material from the original job for this purpose.

Floorboards can be replaced by two methods, either by replacing a rectangular area around the damaged floorboards, or by replacing boards individually.

Removing a rectangular patch is the quickest and easiest method, but it is also more noticeable. This method is recommended when a rug or carpet is going to be placed over the repair area.

I recommend the individual board removal method, which limits the repair to only the boards affected.

To begin this project, use a chisel with the beveled side facing the damaged area, and tap it with a hammer to make vertical score marks across the end of the boards to be removed. Hold the chisel at a 30 degree angle and chip out the board until it is completely cut through (Figure 10.8). The edge of the section not being removed should be sharp and clean.

Be careful not to damage the surrounding boards while chipping out the damaged ones. Cover the good edges with masking tape or a soft cloth for protection.

FIGURE 10.8 Chip out floorboard with a wood chisel at a 30 degree angle.

Now make two horizontal cuts in the remaining board, dividing it into three parts. This will cause the board to split, making it easy to place a pry bar underneath each piece (Figure 10.9). Pry the center part out first, and after it is completely out, pry loose the strip on the groove side of the board, and then finally the strip on the tongue side. Be sure to remove or nail down any exposed nails left by the damaged floorboards.

Use a scrap piece of flooring for a hammering block and tap a cut-to-size replacement board sideways into place. Make sure the groove side goes over the tongue side of the old board.

Next, drill pilot holes at an angle through the tongue of the new board to avoid splitting the wood. Use 8d finishing nails and set them into place through the pilot holes. The final piece will not be able to slide into place like the first strip. You must then remove the lower lip of the groove with a chisel, and then tap the board into place from above (Figure 10.10).

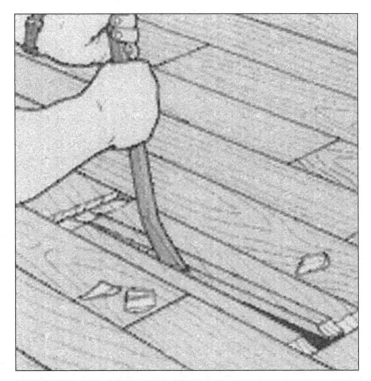

FIGURE 10.9 Split floorboard into three sections.

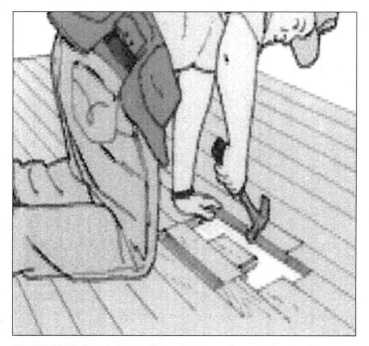

FIGURE 10.10 Tap last board into place.

Set the new floorboards into place with 8d finishing nails through pre-drilled pilot holes. Set the nails flush and fill in with matching wood putty.

Floor repair is a part of floor installations. If you are going to install it, you should know how it is supported, held together, and maintained. The methods of care and maintenance shown in this section cover the basic aspects of a job. There will always be exceptions and new products come out periodically that can revolutionize flooring installation and maintenance.

As you work on each project and gain more experience, you may well be the one to come up with the next advance in flooring.

REFINISHING HARDWOOD FLOORS

Whether in a new or existing home, chances are there is a hardwood floor somewhere. If the house is a new construction, the hardwood floor must be finished out to complete the installation. If the floor already exists, then it may have seen its share of wear and tear over the years, and it is time to bring it back to life. In this chapter we will deal with sanding and finishing out both a new and existing floor.

PREPARING HARDWOOD FLOORS FOR REFINISHING

Basically when we discuss refinishing, we mean sanding and applying a coat of protective finish on top of the hardwood. Simple but necessary things must be done before any sanding and staining can occur. Whether you are working on a new or existing floor, all of these steps should be followed.

Make sure the floor is swept carefully with a soft broom, so as not to scratch the unprotected wood floor. A vacuum cleaner may also be used to get between the cracks of the floor. Again, use only a soft brush attachment when doing so.

Walk over the floor and listen for squeaks. Make sure you fix them before you go to all the work of finishing and sealing the floor.

Go over the floor with your hands and feel for any unevenness. If this can fixed before sanding the floor, do so. Sometimes the sanding process can help make the floor level. If the floor is more than $\frac{1}{32}$ inch higher or lower than other planks, do not use the sander as a method of leveling.

Check for any loose boards or any nails which may be popping up on top of the floorboards. Replace any nails with 6d to 8d flooring nails, or simply nail the existing nails back into the floorboard. Be sure to countersink all of the nails with a nail set and fill with wood putty (Figures 11.1 and 11.2).

Check for large gaps or cracks in the floorboards. Gaps can occur naturally between the floorboards due to humidity and other temperature conditions. Do not try and fill those in, since this may cause cupping or crowning later.

Fill any holes with no-shrink wood putty and let dry. Once it is dry, lightly sand with fine grit sandpaper to a level surface.

After all these items are checked, you may find that there is a board or two that needs to be replaced. Take out the entire board and replace with same type of wood. Nail the new board down in the same manner as original and continue on with project.

Following is a list of tools and materials needed to sand and refinish a floor.

- Floor drum sander
- Edger
- Sanding paper discs (coarse, medium and fine grit)
- Floor polisher
- White or brown buffing pads
- Soft broom
- Vacuum
- Hammer
- 6d to 8d finishing nails
* Nail set
- Paint scraper
- Putty knife
- No-shrink wood filler
- Oscillating electric sander
- Large assortment of rags

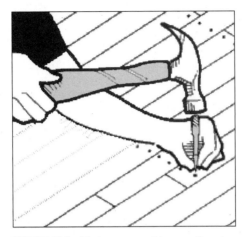

FIGURE 11.1   Replace or re-set loose nails, using a nail set to countersink the nails.

FIGURE 11.2   Fill holes with a color matched non-shrinking wood putty.

SANDING A NEW HARDWOOD FLOOR

Before any sanding begins, make sure to clear the room. If this is a new floor within new construction, there will probably not be any furniture in the room, which will make it easier to work. Next, remove all baseboards around the perimeter of the entire room. Baseboards can be damaged by the drum sander, which is time consuming and costly to replace. Taking the time to remove the baseboards is a cost effective prevention method, and if you do this carefully, you can save the removed baseboards and replace them exactly as they were taken out.

The drum sander and the sandpaper used with it are a key to whether the floor gets an even sanding. If the sander is improperly used, the result may well cause irreparable damage to the floor. Let's first discuss how to properly load a drum sander.

Be certain that you select the grain of sandpaper needed to work with the type of hardwood floor you are refinishing. Make sure the drum sander is unplugged whenever loading sandpaper onto drum. Take a sheet of the sandpaper and put in into the loading slot. Turn the drum a full spin and complete loading the sheet by putting the end in the remaining slot. Do not over-tighten the paper or drum, this may cause stress to the drum and cause chatter marks on the floor. Better to avoid it than go back and try to fix them, because correcting chatter marks is very difficult (Figure 11.3).

Depending on condition of the floor, there are few choices of sandpaper to decide on. Normally a floor will need two types of sandpaper, first the coarser, grittier grade paper to get down past any damage, then the fine paper to smooth it over and give it a clean appearance.

fastfacts

A drum belt sanding machine that has a separate lever for raising and lowering the drum is recommended. The lever will allow for more control while using the sander. This can alleviate a lot of extra problems, like chattering and unevenness that can occur during the sanding process.

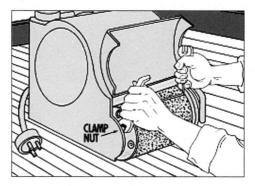

FIGURE 11.3 Placement of sandpaper on roll of drum sander.

Since we are covering new floors for this section, we can use medium grade sandpaper first, then a fine grit to finish up. There should be no damage on a new floor installation, and it will be primarily prepared for the finish coat. Typical grades of sandpaper are noted like this:

- *Coarse Cut*   *36 to 40 grit*
- *Medium Cut*   *50 to 60 grit*
- *Fine Cut*   *80 to 100 grit*

Once the sandpaper is loaded on the belt sander and then adjusted to make level movements over the floor, you're ready to begin the sanding process.

Line the sander up along the right hand wall with about two-thirds length of the floor in front of it. Make a practice run before starting the machine to get a feel for the machine without risking damaging to the floor.

Once the sander is in place, start the machine with the drum raised up off the floor. Make sure to sand the room in the longest direction, always starting at the right hand wall. As you move down the length of the wall, walking slowly and at an even pace, lower the drum until it reaches the floor. As you draw near the end of the strip, gradually raise the drum so it is completely up off the floor by the end of the stretch (Figure 11.4).

Make the same path in the opposite direction. walking backward and using the same method as before. Raise the drum first, then slowly lower it and raise it again as you reach the end.

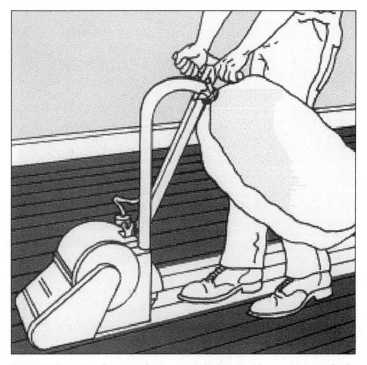

FIGURE 11.4 Run drum sander in a straight line, and be sure you gradually raise the sanding drum before finishing the strip.

When that stretch of floor is completed and you are back at the original starting point, move the machine over about 4 inches. Repeat the same procedure, going forward and backward, raising and lowering the drum sander until each path is complete.

When two-thirds of the room is sanded, turn the sander in the opposite direction and sand it the same way. When the sanding is finished, the two passes should overlap by 2 to 3 feet. This will blend the two areas more naturally than sanding with abrupt stops and starts (Figure 11.5).

Once the first sanding pass has been completed, use a hand sander to go over any areas missed by the drum sander. Areas along the baseboards, corners and closets should be gone over until they match the finish left by the drum sander (Figure 11.6).

FIGURE 11.5   Sand floor by overlapping 2 to 3 feet each time a pass is made.

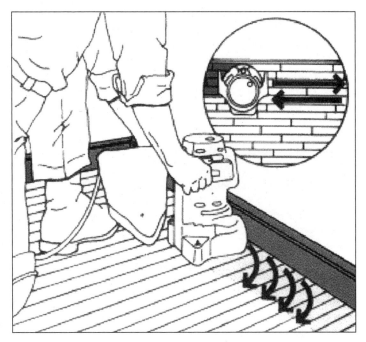

FIGURE 11.6   Use a hand sander in areas that the drum sander cannot reach.

fastfacts

Don't let the sanding drum touch the floor unless the machine is moving forward or backward. If you must stop, gradually raise the drum and stop the machine. If the drum is running and stops over the floor while still in contact with it, you will create a dip in the floor which will be next to impossible to get out.

For new floor sanding, use the next grit of finer sandpaper to smooth out and prepare the hardwood for a clean finish. Any passes after the original sanding will also remove any scratches caused by the sander.

When sanding parquet, block or inlaid wood floor, the direction of the sanding path needs to be adjusted. The first sanding path should cut a diagonal across the grain using medium grit sandpaper. Next use fine grit sandpaper and go in the opposite diagonal direction. Go over the entire floor with a buffer, beginning straight along the longest dimension of the room.

SANDING AN EXISTING FLOOR

An existing hardwood floor can add a few issues to the direction or amount of sanding needed. Unlike a new floor, there are layers of finish and polyurethane to get through, making it necessary to use more aggressive sandpaper grit and multiple directions for the path of the drum sander.

The existing floor is prepared the same way as the preparation of a new floor, making sure the floor is clean, smooth, and flush. The drum sander is loaded the same way with coarse cut or medium cut paper. The type of paper will depend on the damage or amount of finish you need to remove from the existing floor.

When using different levels of sandpaper, do not skip more than one grit type. In other words, don't go from a coarse grit to a fine grit, go from coarse, to medium, and then to fine. If the floor is cupped or crowned, it may be necessary to make the first path at a 45 degree angle rather than straight along the longest edge of the floor. All paths made after the first 45 degree angle should be parallel to the direction of the flooring. Both paths should use the same grit sandpaper.

The more coarse sandpaper can also be used as a leveler for the floor if needed. Unlike new floors, older hardwood flooring like oak is thicker and can be refinished many times without going too deep. If thinner solid oak floors are sanded too much, they can wear down to the groove edge and reveal nails, or even cause breakage.

For more persistent scratches or damage, make multiple angle paths forward and backward over the affected area. Start the first path at a 30 degree angle to the scratch or damage. Be sure to work the damaged area from the front and back to avoid making it appear larger. Once the damaged area has been sanded down with the belt sander, sand the area with the grain of the wood to get rid of all heavy sanding marks.

If an edger is being used to complete the hard to get areas, be sure that you don't press down to get deeper into the finish or scratches. This will only cause ridges and grooves that will be tough to remove later. Always use a fine grit sandpaper with the edger. The edger will sand a smaller surface than the drum sander, but it is much more agile and detail-oriented.

For the really difficult areas, hand-scraping may be necessary. Any marks made with an orbital sander or a hand scraper may be seen once a finish is applied to a floor. It is best to hand sand any areas with fine grit sandpaper after any marks have been made.

Sometimes, especially with any common grade wood floors filled with many cracks and gaps, it is necessary to trowel fill an entire floor. Let the filler dry before the final two paths of sanding are performed. Filler will typically dry overnight.

A total of three paths of sanding, followed by discing, which can blend all the layers of sanding are recommended before applying a finish to a floor (Figure 11.7). If you use only two levels of sanding, the finished surface will have a coarser final appearance.

FINISHING THE HARDWOOD FLOOR

Now that both new and existing floor sanding have been discussed, it is time to finish the floor. Wood flooring is treated and finished the same way, no matter if it is old or new. At this point, the floors have been sanded and smoothed out, and are now all basically the same surface.

Once a floor has been sanded, it should be finished as quickly as possible to protect it from traffic and stains that can become permanent

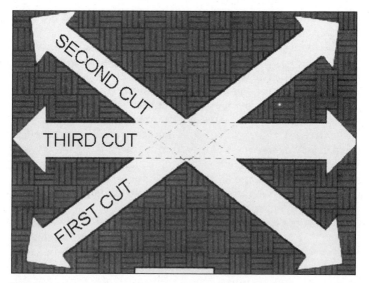

FIGURE 11.7 Use a total of three paths for sanding the floor.

if left untreated. Make sure that sanded floors have ample time for the grain to settle. Skipping this step can create a rough surface which will make for additional work during the finishing process. If a rough surface appears, it can be rubbed down with fine steel wool until the roughness flattens out.

FINISH MATERIALS

Whether a floor is stained and polyurethaned, or just sealed for a finish, there are several products to choose from. The most important factor to remember is that most of these finishes are very toxic, and proper precautions must be adhered to.

- *Always ventilate the room where the product is being used*
- *Read and follow the manufacturers' instructions listed on the product label*
- *Some products require that the user wear a mask and or gloves. If the instructions indicate doing so, do it. Don't endanger your health.*
- *When you have finished using the product, save any extra for future touch-up or save the name of the color and the manufacturer*

- *Dispose of any remaining product that will not be used safely and properly*
- *Most municipalities have restrictions for dumping when it comes to paint and other toxins. Be sure you understand these restrictions before disposing of any chemical products.*

Finish materials can be grouped into three general types:

- *Sealers*
- *Stain or sealer combinations*
- *Fast-drying stains and sealers*

Penetrating sealers do just what they claim; they soak into the wooden floor and harden to create a seal. This type of finish is the easiest to apply and maintain. The sealer wears away as the wood wears and will not chip or scratch. After a long period of wear and tear, floors with penetrating sealers can easily be restored with another thin coat of sealer. In between restorations, a thin coat of wax is all that is necessary to maintain the wood. Exceptionally worn areas can be refinished without showing lap marks when a new finish is applied over the top. Some penetrating sealers can contain stain and color and seal the floor all at the same time.

Stain sealer combinations are great if you want to stain and seal a floor at the same time. Combination stain sealers takes more time to dry than if you apply a coat of stain first, let it dry, and then apply a sealer. Combination sealers can take at least 48 hours or more to dry, but there is only the one application process.

Stain applications are great if you want the job to dry relatively quickly. Most stains will dry in about eight hours, and then a sealer can be applied and allowed to dry in another eight hours. Stains can have a tendency to overlap and create uneven color.

Fat-drying sealers are for the highly experienced finisher. These products dry almost as quickly as they are applied to the floor. There is not a lot of time to make or fix mistakes with these products. The drawbacks are the overlapping which can happen if the product is not applied correctly.

APPLICATION OF FINISHES

Penetrating stain sealers can be applied with rags, simply by wiping it as even as possible over the floor. Begin in one corner and work your way out of the room, following the direction of the floor. Use a

fastfacts

Be aware that when a bare hand or other part of the body comes in contact with a finished or unfinished piece of flooring, that area can become discolored. Skin contains all sorts of oil which will transfer to the floor and be absorbed, just like water. Wear gloves or shoes when maneuvering around the floor before, during, and after application of stains and sealers.

paint brush along the edge of the floor to avoid getting sealer on the baseboards.

The longer the stain sealer is left on, the darker it will become. If you wipe it off right away it will still leave enough product on the floor to make a noticeable difference. Do not let the stain sealer dry before wiping it off. This could cause unevenness in color.

Once the stain sealer is dry, use the floor polisher to buff the floor with a fine steel wool disc or brown fiber buffer. Clean any dust with a vacuum or broom. A wax can then be applied with the floor polisher by buffing it to a satin sheen. An amount of wax about the size of a walnut is all that is needed to buff and protect an average size room.

SURFACE FINISHES

Polyurethane, water-based, and cured urethanes are all considered durable, water-proof finishes. These finishes remain on the surface and protect the floor from stains and wear and tear. They are available in high-gloss, semi-gloss, satin and matte finishes.

Oil-modified polyurethane is the most common of the surface finishes. It is the easiest to apply and can age to a little deeper patina than its original finish.

Water based finishes are clear, durable, non-yellowing, and non-flammable at the time of application. They are similar in ease and care to the oil-modified products, but offer an odor free, quick application.

If these products are applied after a sealer or stain, make sure the floor is completely dry before application. Always follow up with a steel wool or fiber cloth buffing of the floor and then vacuuming of

fastfacts

If water based urethane is to be applied to a floor, do not use the steel wool buffing method. The steel wool fibers will rust on contact with the water and will discolor any finish. Always use the fiber cloth buffing method.

any residue. Oil modified polyurethanes are highly toxic and should be treated with great care. Turn off any open flames in the room, including pilot lights.

This product can be applied with a soft brush after being poured into a metal bucket or tray. Apply in the direction of the floor grain and be sure to overlap any areas while they are still wet. Do not over saturate any area. Over saturation may cause bubbling on the surface. Allow the finish to dry overnight before walking over it. Clean off any dust residue with a vacuum or soft broom. A second or third coat may be applied if extra durability and protection are desired. Be sure to buff between each layer with fine steel wool.

Water-based urethanes are applied a little differently than oil modified urethanes. Water based urethanes are poured directly onto the floor along the starting wall. A small, thin trail should be started about 4 to 5 inches away from the edge of the baseboard. Then use the applicator and spread the urethane out across the floor. Once you run out of urethane, pour a little more out onto floor in the same manner and spread accordingly. Make sure all doors and win-

fastfacts

A good test to find out what type of surface the old floor has is to take a sharp object and, in an inconspicuous place, make a small scratch. If the finish does not flake off, a penetrating seal was probably used. If the finish flakes off, a surface finish was probably used. These tests will help determine which type of product to refinish the floor with correctly.

dows are open for ventilation, allowing for a quick drying time. A second coat can usually be applied the same day as the first. Be sure to buff between each coat with a fiber cloth.

Avoid waxing a hardwood floor that has already been waxed. Apply wax only after the floor has been stripped and sanded, or if the finish has been worn off.

Hardwood floors are the longest lasting floors that can be installed. If they are maintained over time, chances are they will never have to be re-finished. If refinishing is necessary, follow these guidelines to make the job simple and easy.

12

WORKSITE SAFETY

Worksite safety doesn't get as much attention as it should. Far too many people are injured on jobs every year. Most of the injuries could be prevented, but they are not. Why is this? People are in a hurry to make a few extra bucks, so they cut corners. This happens with flooring contractors and piece workers. It even affects hourly installers who what to shave 15 minutes off their workday, so that they can head back to the shop early (Figure 12.1).

Based on field experience, most accidents occur as a result of negligence. Installers try to cut a corner, and they wind up getting hurt. Accidents can be completely avoided if common sense and awareness prevail. Sometimes you don't get a second chance, and the life you affect may not be your own. So, let's look at some sensible safety procedures that you can implement in your daily activity.

VERY DANGEROUS

Floor installations can be a dangerous trade. The tools of the trade have the potential to be killers. Requirements of the job can place you in positions where a lack of concentration could result in serious injury or death. The fact that floor installation can be dangerous is no reason to rule out the trade as your profession. Driving can be extremely dangerous, but few people never set foot in an automobile out of fear.

General Safe Working Habits

1. Wear safety equipment.

2. Observe all safety rules at the particular location.

3. Be aware of any potential dangers in the specific situation.

4. Keep tools in good condition.

FIGURE 12.1   General safe working habits.

Fear is generally a result of ignorance. When you have a depth of knowledge and skill, fear begins to subside. As you become more accomplished at what you do, fear is forgotten. While it is advisable to learn to work without fear, you should never work without respect. There is a huge difference between fear and respect.

You must respect the position you are putting yourself in.

Many young installers are fearless in the beginning. They think nothing of darting around on a balcony or jumping down a flight of steps. As their careers progress, they usually hear about or see on-the-job accidents. Someone gets badly cut with a power saw. Somebody falls off a landing. A careless installer steps into a flooded basement and is electrocuted because of submerged equipment. The list of possible job-related injuries is a long one.

Millions of people are hurt every year in job-related accidents. Most of these people were not following solid safety procedures. Sure, some of them were victims of unavoidable accidents, but most were hurt by their own hand, in one way or another. You don't have to be one of these statistics (Figure 12.2)

As an installer, you will be doing some dangerous work. You will be drilling holes, running sanding machines, cutting baseboards and door trim, and a lot of other potentially dangerous jobs. Hopefully, your employer will provide you with quality tools and equipment. If you have the right tool for the job, you are off to a good start in staying safe.

Safety training is another factor you should seek from your employer. Some flooring contractors fail to tell their employees how to do their jobs safely. It is easy for someone, like an experienced

Safe Dressing Habits

1. Do not wear clothing that can be ignited easily.

2. Do not wear loose clothing, wide sleeves, ties or jewelry (bracelets, necklaces) that can become caught in a tool or otherwise interfere with work. This caution is especially important when working with electrical machinery.

3. Wear gloves to handle hot or cold pipes and fittings.

4. Wear heavy-duty boots. Avoid wearing sneakers on the job. Nails can easily penetrate sneakers and cause a serious injury (especially if the nail is rusty).

5. Always tighten shoelaces. Loose shoelaces can easily cause you to fall, possibly leading to injury to yourself or other workers.

6. Wear a hard hat on major construction sites to protect the head from falling objects.

FIGURE 12.2   Safe dressing habits.

installer, who knows a job inside and out to forget to inform an inexperienced person of potential danger.

For example, an installer might tell you to break up the stone tile in a room to allow the installation of a new floor and never consider telling you to wear safety glasses. The installer will assume you know the stone is going to fly up in your face as it is chiseled up. However, as a rookie, you might not know about the reaction stone has when hit with a cold chisel. One swing of the hammer could cause extreme damage to your eyesight.

Simple jobs, like the one in the example, are all it takes to ruin a career. You might be really on your toes when asked to cut door trim with a power saw, but how much thought are you going to give to driving in a few finishing nails? The risk of running the power saw is obvious. Smashing your fingers with a hammer is not so obvious. Either way, you can have a work-stopping injury.

Safety is a serious issue. Some job sites are very strict in the safety requirements maintained. But a lot of jobs have no written rules of safety. If you are working on a commercial job, supervisors are likely to make sure you abide by the rules of the Occupational Safety and

Health Administration (OSHA). Failure to comply with OSHA regulations can result in stiff financial penalties. However, if you are working in residential flooring, you may never set foot on a job where OSHA regulations are observed.

In all cases, you are responsible for your own safety. Your employer and OSHA can help you to remain safe, but in the end, it is up to you. You are the one who has to know what to do and how to do it. And not only do you have to take responsibility for your own actions, you also have to watch out for the actions of others. It is not unlikely that you could be injured by someone else's carelessness. Now that you have had the primer course, let's get down to the specifics of job-related safety (Figure 12.3)

As we move into specifics, you will find the suggestions in this chapter broken down into various categories. Each category will deal with specific safety issues related to the category. For example, in the section on tool safety, you will learn procedures for working safely with tools. As you move from section to section, you may notice some overlapping of safety tips. For example, in the section on general safety, you will see that it is wise to work without wearing jewelry. Then you will find jewelry mentioned again in the tool section. The duplication is done to pinpoint definite safety risks and procedures. We will start into the various sections with general safety.

GENERAL SAFETY

General safety covers a lot of territory. It starts from the time you get into the company vehicle and carries you right through to the end of the day. Much of the general safety recommendations involve the use of common sense. Now, let's get started.

Vehicles

Many installers are given company trucks for their use in getting to and from jobs. You will probably spend a lot of time loading and unloading company trucks. And, of course, you will spend time either riding in or driving them. All of these areas can threaten your safety.

If you will be driving the truck, take the time to get used to how it handles. Loaded delivery trucks don't drive like the family car. Remember to check the vehicle's fluids, tires, lights, and related equipment. Many flooring trucks are old and have seen better days.

Safe Operation of Grinders

1. Read the operating instructions before starting to use the grinder.

2. Do not wear any loose clothing or jewelry.

3. Wear safety glasses or goggles.

4. Do not wear gloves while using the machine.

5. Shut the machine off promptly when you are finished using it.

FIGURE 12.3 Safe operation of grinders.

Failure to check the vehicle's equipment could result in unwanted headaches. Also, remember to use the seat belts—they do save lives.

Apprentices are normally charged with the duty of unloading the truck at the job site. There are a lot of ways to get hurt in doing this job. Many flooring trucks use roof racks to haul floorboards and ladders. If you are unloading these items, make sure they will not come into contact with low-hanging electrical wires. If you are unloading heavy items, don't put your body in awkward positions. Learn the proper ways for lifting, and never lift objects inappropriately. If the weather is wet, be careful climbing on the truck. Step bumpers get slippery, and a fall can impale you on an object or bang up your knee.

When it is time to load the truck, observe the same safety precautions you did in unloading. In addition to these considerations, always make sure your load is packed evenly and well secured. Be especially careful of any load you attach to the roof rack, and always double check the cargo doors on trucks with utility bodies.

It will not only be embarrassing to lose your load going down the road, it could be deadly. I have seen load of hardwood flooring fly out of the back of a pick-up truck as the truck was rolling up an interstate highway. As a young helper, I lost a load of carpet in the middle of a busy intersection. In that same year, the cargo doors on the utility body of my truck flew open as I came off a ramp, onto a major highway. Tools scattered across two lanes of traffic. These types of accidents don't have to happen. It's your job to make sure they don't.

CLOTHING

Clothing is responsible for a lot of on-the-job injuries. Sometimes it is the lack of clothing that causes the accidents, and there are many times when too much clothing creates the problem. Generally, it is wise not to wear loose fitting clothes. Shirttails should be tucked in, and short-sleeve shirts are safer than long-sleeved shirts when operating some types of equipment.

Caps can save you from minor inconveniences, like getting glue in your hair, and hard hats provide some protection from potentially damaging accidents, like having a ceramic tile dropped on your head. If you have long hair, keep it up and under a hat.

Good footwear is essential in the trade. Normally a strong pair of hunting-style boots will be best. The thick soles provide some protection from nails and other sharp objects you may step on. Boots with steel toes can make a big difference in your physical well being. If you are going to be climbing, wear foot gear with a flexible sole that grips well. Gloves can keep your hands warm and clean, but they can also contribute to serious accidents. Wear gloves sparingly, depending upon the job you are doing.

JEWELRY

On the whole, jewelry should not be worn in the workplace. Rings can inflict deep cuts in your fingers. They can also work with machinery to amputate fingers. Chains and bracelets are equally dangerous, probably more so.

EYE AND EAR PROTECTION

Eye and ear protection is often overlooked. An inexpensive pair of safety glasses can prevent you from spending the rest of your life blind. Ear protection reduces the effect of loud noises, such as power saws and drills. You may not notice much benefit now, but in later years you will be glad you wore it. If you don't want to lose your hearing, wear ear protection when subjected to loud noises.

PADS

Kneepads not only make an installer's job more comfortable, they help to protect the knees. Most installers spend a lot of time on their knees, and pads should be worn to ensure that they can continue to work for many years.

All too many people believe that working without safety glasses, ear protection, and so forth makes them tough. That's just not true. It may make them appear dumb, and it may get them hurt, but it does not make them look tough. If anything, it makes them look stupid or inexperienced.

Don't fall into the trap so many young installers do. Never let people goad you into bad safety practices. Some installers are going to laugh at your earplugs. Let them laugh, you will still have good hearing when they are shopping for hearing aids. I'm dead serious about this issue. There is nothing sissy about safety. Wear your gear in confidence, and don't let the few jokesters get to you.

TOOL SAFETY

Tool safety is a big issue in floor installation. Anyone in the flooring trade will work with numerous tools. All of these tools are potentially dangerous, but some of them are especially hazardous. This section is broken down by the various tools used on the job. You cannot afford to start working without the basics in tool safety. The more you can absorb on tool safety, the better off you will be (Figure 12.4).

The best starting point is reading all literature available from the manufacturers of your tools. The people that make the tools provide some good safety suggestions with them. Read and follow the manufacturers' recommendations.

The next step in working safely with your tools is to ask questions. If you don't understand how a tool operates, ask someone to explain it to you. Don't experiment on your own; the price you pay could be much too high (Figure 12.5).

Common sense is irreplaceable in the safe operation of tools. If you see an electrical cord with cut insulation, you should have enough common sense to avoid using it. In addition to this type of simple observation, you will learn some interesting facts about tool safety.

There are some basic principals to apply to all of your work with tools. We will start with the basics, and then we will move on to specific tools. Here are the basics:

- *Keep body parts away from moving parts*
- *Don't work with poor lighting conditions*
- *Be careful of wet areas when working with electrical tools*
- *If special clothing is recommended for working with your tools, wear it*
- *Use tools only for their intended purposes*
- *Get to know your tools well*
- *Keep your tools in good condition*

Now, let's take a close look at the tools you are likely to use. Floor installers use a wide variety of hand tools and electrical tools. They also use specialty tools. So let's see how you can use all these tools without injury.

Drills and Bits

Not all drills used by installers are the little pistol-grip, hand-held types most people think of. Some jobs require the use of large, powerful right-angle drills. These drills have enormous power when they

---

Safe Use of Hand Tools

1. Use the right tool for the job.

2. Read any instructions that come with the tool unless you are thoroughly familiar with its use.

3. Wipe and clean all tools after each use. If any other cleaning is necessary, do it periodically.

4. Keep tools in good condition. Chisels should be kept sharp and any mushroomed heads kept ground smooth; saw blades should be kept sharp; pipe wrenches should be kept free of debris and the teeth kept clean; etc.

5. Do not carry small tools in your pocket, especially when working on a ladder or scaffolding. If you should fall, the tools might penetrate your body and cause serious injury.

FIGURE 12.4 Safe use of hand tools.

Safe Use of Electric Tools

1. Always use a three-prong plug with an electric tool.

2. Read all instructions concerning the use of the tool (unless you are thoroughly familiar with its use).

3. Make sure that all electrical equipment is properly grounded. Ground fault circuit interrupters (GFCI) are required by OSHA regulations in many situations.

4. Use proper-sized extension cords. (Undersized wires can burn out a motor, cause damage to the equipment, and present a hazardous situation.

5. Never run an extension cord through water or through any area where it can be cut, kinked, or run over by machinery.

6. Always hook up an extension cord to the equipment and then plug it into the main electrical outlet—not vice versa.

7. Coil up and store extension cords in a dry area.

FIGURE 12.5 Safe uses of electric tools.

get in a bind. Hitting a nail or a knot in the wood being drilled can do a lot of damage. You can break fingers, lose teeth, suffer head injuries, and a lot more. As with all electrical tools, you should always check the electrical cord before using your drill. If the cord is not in good shape, don't use the drill.

Always know what you are drilling into. If you are doing new-construction work it is fairly easy to look before you drill. However, drilling in a remodeling job can be much more difficult. You cannot always see what you are getting into. If you are unfortunate enough to drill into a hot wire, you can get a considerable electrical shock.

The bits you use in a drill are part of the safe operation of the tool. If your drill bits are dull, sharpen them. Dull bits are much more dangerous than sharp ones. When you are using a standard bit to drill through thin wood, like plywood, be careful. Once the worm driver of the bit penetrates the plywood fully, the teeth on the bit can bite and jump, causing you to lose control of the drill. If you will be drilling metal, be aware that the metal shavings will be sharp and hot.

Power Saws

Installers don't use power saws as much as carpenters, but they do use them. The most common type of saw used by installers is the power saw. These saws are used to cut door trim, and a whole lot more. In addition to power saws, installers use circular saws and chop saws. All of the saws have the potential for serious injury.

Reciprocating saws are reasonably safe. Most models are insulated to help avoid electrical shocks if a hot wire is cut. The blade is typically a safe distance from the user, and the saws are pretty easy to hold and control. However, the brittle blades do break. This could result in an eye injury.

Wet saws are used to cut ceramic and stone tile. Eye protection is an absolute necessity when using these saws. Flying chips of stone or ceramic can do irreparable harm to your eyesight. Many intricate cuts with these saws involve small pieces of tile, which means your fingers may be close to the spinning blade. Be sure you set the saw up in a secure location, out of the way of foot traffic. Don't allow interruptions while you are making your cuts. A minor distraction can result in a major injury.

Circular saws are often used by installers. The blades on these saws can bind and cause the saws to kick back. Chop saws are commonly used to cut hardwood. Keep your body parts out of the way and always wear eye protection.

Sanding Machines

Sanding machines are often used for installing or refinishing hardwood floors. Great care must be taken when installing the sandpaper on the belt of the drum sander. If the sandpaper is not properly installed, it can fly off and injure the operator or anyone else in the room. Obviously, the wearing of safety glasses is a good thing when operating the sander.

When the sanding machine is in operation, a large amount of dust and wood particles will be discharged into the air, so along with the protective eyewear a mask is also a good idea.

Just think over the years how much dust and particles are absorbed by the body on average. Wearing the mask can extend a life and the length of time one works.

A sanding machine is a large, sometimes unruly device which must be controlled. Before releasing it on the job, make sure all cords and wires are removed from its path. Also, make sure the floor is not wet. Even a grounded electrical device can cause a fire or become inactive due to excessive amounts of water. Also, be sure the sanding path is clear of nail, staples and other debris which can be sent flying off the sander.

Air-Powered Tools

Air-powered tools are typically used when installing carpet or hardwood with a power nail gun or staple gun. Make sure the guns are properly loaded so they do not go off unless you want them to.

Air-powered tools create a great deal of force when released. Do not plug them in until you are ready to use them. Wear safety glasses in case of misfires which could backfire a nail or staple into your eye or other areas of the body.

Always aim the gun at the job at hand. Never pretend to shoot a co-worker when the gun is loaded and plugged in. Imagine it as real gun and what the results would be.

Powder-Actuated Tools

Powder-actuated tools are used by installers to secure objects to hard surfaces, like concrete. If the user is properly trained, these tools are not too dangerous. However, good training, eye protection, and ear protection are all necessary. Misfires and chipping hard surfaces are the most common problem with these tools.

Screwdrivers and Chisels

Eye injuries and puncture wounds are common when working with screwdrivers and chisels. When the tools are use properly and safety glasses are worn, few accidents occur.

The key to avoiding injury with most hand tools is simply to use the right tool for the job. If you use a wrench as a hammer or a screwdriver as a chisel, you are asking for trouble.

There are, of course, other types of tools and safety hazards found in the flooring trade. However, this list covers the ones that result in the most injuries. In all cases, observe proper safety procedures and utilize safety gear, such as eye and ear protection.

CO-WORKER SAFETY

Co-worker safety is the last segment of this chapter. I am including it because workers are frequently injured by the actions of co-workers. This section is meant to protect you from others and to make you aware of how your actions might affect your co-workers.

Most installers find themselves working around other people. This is especially true on construction jobs. When working around other people, you must be aware of their actions, as well as your own. If you are walking out of a house to get something off the truck and a roll of roofing paper gets away from a roofer, you could get an instant headache.

If you don't pay attention to what is going on around you, it is possible to wind up in all sorts of trouble. Cranes lose their loads some times, and such a load landing on you is likely to be fatal. Equipment operators don't always see the installer kneeling down for a piece of tile. It's not hard to have a close encounter with heavy equipment.

You have to watch out for yourself at all times. As you gain field experience, you will develop a second nature for impending co-worker problems. You will learn to sense when something is wrong or is about to go wrong. But you have to stay alive and healthy long enough to get that experience.

Always be aware of what is going on over your head. Avoid working under other people and hazardous overhead conditions. Let people know where you are, so you won't get stranded on a roof or in an attic when your ladder is moved or falls over.

You must also remember that your actions could harm co-workers. If you are on a roof to flash a pipe and your hammer gets away from you, somebody could get hurt. Open communication between workers is one of the best ways to avoid injuries. If everyone knows where everyone else is working, injuries are less likely. Primarily, think and think some more and remain alert at all times. There Is no substitute for common sense.

13

FIRST AID

Everyone should invest some time in learning the basics of first aid. You never know when having skills in first aid treatments may save your life. Installers live what can be a dangerous life. On the job injuries are not uncommon. Most injuries are fairly minor, but they often require treatment. Do you know the right way to get a sliver of ceramic tile out of your hand? If your helper suffers from an electrical shock when a drill cord goes bad, do you know what to do? Well, many installers don't possess good first aid skills.

Before we get too far into this chapter, there are a few points are want to make. First of all, I'm not a medical doctor or any type of trained medical-care person. I've taken first aid classes, but I'm certainly not an authority on medical issues. The suggestions that I will give you in this chapter are for informational purposes only. This book is not a substitute for first aid training offered by qualified professionals.

My intent here is to make you aware of some basic first aid procedures that can make life on the job much easier. But, I want you to understand that I'm not advising you to use my advice to administer first aid. Hopefully, this chapter will show you the types of advantages you can gain from taking first aid classes. Before you attempt first aid on anyone, including yourself, you should attend a structured, approved first aid class. I'm going to give you information that is as accurate as I can make it, but don't assume that my words are enough. Take a little time to seek professional training in the art of first aid. You may never use what you learn, but the one time it is needed, you will be glad you made the effort to learn what to do. With this said, let's jump right into some tips on first aid.

OPEN WOUNDS

Open wounds are a common problem for installers. Many tools and materials used by installers can create open wounds. What should you do if you or one of your workers is cut?

* *Stop the bleeding as soon as possible*
* *Disinfect and protect the wound from contamination*
* *You may have to take steps to avoid shock symptoms*
* *Once the patient is stable, seek medical attention for severe cuts*

When a bad cut is encountered, the victim may slip into shock. A loss of consciousness could result from a loss of blood. Death from extreme bleeding is also a risk. As a first aid provider, you must act quickly to reduce the risk of serious complications.

Bleeding

To stop bleeding, direct pressure is normally a good tactic. This may be as crude as clamping your hand over the wound, but a cleaner compression is desirable. Ideally, a sterile material should be placed over the wound and secured, normally with tape (even if it's duct tape). Whenever possible, wear rubber gloves to protect yourself from possible disease transfer if you are working on someone else. Thick gauze used as a pressure material can absorb blood and begin to allow the clotting process to start.

Bad wounds may bleed right through the compress material. If this happens, don't remove the blood-soaked material. Add a new layer of material over it. Keep pressure on the wound. If you are not prepared with a first aid kit, you could substitute gauze and tape with strips cut from clothing that can be tied in place over the wound.

When you are dealing with a bleeding wound, it is usually best to elevate it. If you suspect a fractured or broken bone in the area of the wound, elevation may not be practical. When we talk about elevating a wound, it simply means to raise the wound above the level of the victim's heart. This helps the blood flow to slow, due to gravity.

Super Serious

Super serious bleeding might not stop even after a compression bandage is applied and the wound is elevated. When this is the case, you must resort to putting pressure on the main artery which is producing the blood. Constricting an artery is not an alternative for the steps that we have discussed previously.

Putting pressure on an artery is serious business. First, you must be able to locate the artery, and you should not keep the artery constricted any longer than necessary. You may have to apply pressure for awhile, release it, and then apply it again. It's important that you do not restrict the flow of blood in arteries for long periods of time. I hesitate to go into too much detail on this process, as I feel it is a method that you should be taught in a controlled, classroom situation. However, I will hit the high spots. But remember, these words are not a substitute for professional training from qualified instructors.

Open arm wounds are controlled with the brachial artery. The location of this artery is in the area between the biceps and triceps, on the inside of the arm. It's about halfway between the armpit and the elbow. Pressure is created with the flat parts of your fingertips. Basically, you are holding the victim's wrist with one hand and closing off the artery with your other hand. Pressure exerted by your fingers pushes the artery against the arm bone and restricts blood flow. Again, don't attempt this type of first aid until you have been trained properly in the execution of the procedure.

Severe leg wounds may require the constriction of the femoral artery. This artery is located in the pelvic region. Normally, bleeding victims are placed on their backs for this procedure. The heel of a hand is placed on the artery to restrict blood flow. In some cases, fingertips are used to apply pressure. I'm uncomfortable with going into great detail on these procedures, because I don't want you to rely solely on what I'm telling you. It's enough that you understand that knowing when and where to apply pressure to arteries can save lives and that you should seek professional training in these techniques.

Tourniquets

Tourniquets get a lot of attention in movies, but they can do as much harm as they do good if not used properly. A tourniquet should only be used in a life-threatening situation. When a tourniquet is applied, there is a risk of losing the limb to which the restriction is applied. This is obviously a serious decision and one that must be made only when all other means of stopping blood loss have been exhausted.

Unfortunately, installers might run into a situation where a tourniquet is the only answer. For example, if a worker allowed a power saw to get out of control, a hand might be severed or some other type of life-threatening injury could occur. This would be cause for the use of a tourniquet. Let me give you a basic overview of what's involved when a tourniquet is used.

Tourniquets should be at least two inches wide. A tourniquet should be placed at a point that is above a wound, between the bleeding and the victim's heart. However, the binding should not encroach directly on the wound area. Tourniquets can be fashioned out of many materials. If you are using strips of cloth, wrap the cloth around the limb that is wounded and tie a knot in the material. Use a stick, screwdriver, or whatever else you can lay your hands on to tighten the binding.

Once you have made a commitment to apply a tourniquet, the wrapping should be removed only by a physician. It's a good idea to note the time that a tourniquet is applied, as this will help doctors later in assessing their options. As an extension of the tourniquet treatment, you will most likely have to treat the patient for shock.

Infection

Infection is always a concern with open wounds. When a wound is serious enough to require a compression bandage, don't attempt to clean the cut. Keep pressure on the wound to stop bleeding. In cases of sever wounds, be on the look out for shock symptoms, and be prepared to treat them. Your primary concern with a serious open wound is to stop the bleeding and gain professional medical help as soon as possible.

Lesser cuts, which are more common than deep ones, should be cleaned. Regular soap and water can be used to clean a wound before applying a bandage. Remember, we are talking about minor cuts and scrapes at this point. Flush the wound generously with clean water. A piece of sterile gauze can be used to pat the wound dry. Then a clean, dry bandage can be applied to protect the wound while in transport to a medical facility.

SPLINTERS AND SUCH

Splinters and such foreign objects often invade the skin of installers. Getting these items out cleanly is best done by a doctor, but there

fastfacts

- *Use direct pressure to stop bleeding*
- *Wear rubber gloves to prevent direct contact with a victim's blood*
- *When feasible, elevate the part of the body that is bleeding*
- *Extremely serious bleeding can require you to put pressure on the artery supplying the blood to the wound area*
- *Tourniquets can do more harm than good*
- *Tourniquets should be at least two inches wide*
- *Tourniquets should be placed above the bleeding wound, between the bleeding and the victim's heart*
- *Tourniquets should not be applied directly on the wound area*
- *Tourniquets should only be removed by trained medical professionals*
- *If you apply a tourniquet, note the time that you apply the tourniquet*
- *When a bleeding wound requires a compression bandage, don't attempt to clean the wound. Simply apply compression quickly*
- *Watch victims with serious bleeding for symptoms of shock*
- *Lesser bleeding wounds should be cleaned before being bandaged*

are some on the job methods that you might want to try. A magnifying glass and a pair of tweezers work well together when removing embedded objects, such as splinters and slivers of tile. Ideally, tweezers being used should be sterilized either over an open flame, such as the flame of your torch, or in boiling water.

Splinters and slivers that are submerged beneath the skin can often be lifted out with the tip of a sterilized needle. The use of a needle in conjunction with a pair of tweezers is very effective in the removal of most simple splinters. If you are dealing with something that has gone extremely deep into tissue, it is best to leave the object alone until a doctor can remove it.

EYE INJURIES

Eye injuries are very common on construction and remodeling jobs. Most of these injuries could be avoided if proper eye protection was worn, but far too many workers don't wear safety glasses and goggles. This sets the stage for eye irritations and injuries.

Before you attempt to help someone who is suffering from an eye injury, you should wash your hands thoroughly. I know this is not always possible on construction sites, but cleaning your hands is advantageous. In the meantime, keep the victim from rubbing the injured eye. Rubbing can make matters much worse.

Never attempt to remove a foreign object from someone's eye with the use of a rigid device, such as a toothpick. Cotton swabs that have been wetted can serve well as a magnet to remove some types of invasion objects. If the person you are helping has something embedded in an eye, get the person to a doctor as soon as possible. Don't attempt to remove the object yourself.

When you are investigating the cause of an eye injury, you should pull down the lower lid of the eye to determine if you can see the object causing trouble. A floating object, such as a piece of sawdust trapped between an eye and an eyelid can be removed with a tissue, a damp cotton swab, or even a clean handkerchief. Don't allow dry cotton material to come into contact with an eye.

If looking under the lower lid doesn't a source of discomfort, check under the lower lid. Clean water can be used to flush out many eye contaminants without much risk of damage to the eye. Objects that cannot be removed easily should be left alone until a physician can take over.

- *Wash your hands, if possible, before treating eye injuries*
- *Don't rub an eye wound*
- *Don't attempt to remove embedded items from an eye*
- *Clean water can be used to flush out some eye irritants*

SCALP INJURIES

Scalp injuries can be misleading. What looks like a serious wound can be a fairly minor cut. On the other hand, what appears to be only a cut can involve a fractured skull. If you or someone around you sustains a scalp injury, such as having a hammer fall on your head from an overhead worker, take it seriously. Don't attempt to clean the wound. Expect profuse bleeding.

If you don't suspect a skull fracture, raise the victim's head and shoulders to reduce bleeding. Try not to bend the neck. Put a sterile bandage over the wound, but don't apply excessive pressure. If there is a bone fracture, pressure could worsen the situation. Secure the bandage with gauze or some other material that you can wrap around it. Seek medical attention immediately.

FACIAL INJURIES

Facial injuries can occur on flooring jobs. I've seen helpers let their drills get away from them with the result being hard knocks to the face. On one occasion, I remember a tooth being lost, and split lips and tongues that have been bitten are common when a drill goes on a rampage.

Extremely bad facial injuries can cause a blockage of the victim's air passages. This, of course, is a very serious condition. It's critical that air passages be open at all times. If the person's mouth contains broken teeth or dentures, remove them. Be careful not to jar the individual's spine if you have reason to believe there may be injury to the back or neck.

Conscious victims should be positioned, when possible, so that secretions from the mouth and nose will drain out. Shock is a potential concern in severe facial injuries. For most on the job injuries, installers should be treated for comfort and sent for medical attention.

NOSE BLEEDS

Nose bleeds are not usually difficult to treat. Typically, pressure applied to the side of the nose where bleeding is occurring will stop the flow of blood. Applying cold compresses can also help. If external pressure is not stopping the bleeding, use a small, clean pad of gauze to create a dam on the inside of the nose. Then, apply pressure on the outside of the nose. This will almost always work. If it doesn't, get to a doctor.

BACK INJURIES

There is really only one thing that you need to know about back injuries. Don't move the injured party. Call for professional help and see that the victim remains still until help arrives. Moving someone who has suffered a back injury can be very risky. Don't do it unless there is a life-threatening cause for your action, such as a person trapped in a fire or some other type of deadly situation.

LEGS AND FEET

Legs and feet sometimes become injured on job sites. When someone suffers a minor foot or leg injury, you should clean and cover the wound. Bandages should be supportive without being constrictive. The appendage should be elevated above the victim's heart level when possible. Prohibit the person from walking. Remove boots and socks so that you can keep an eye on the person's toes. If the toes begin to swell or turn blue, loosen the supportive bandages.

Blisters

Blisters may not seem like much of an emergency, but they can sure take the steam out of a helper or installer. In most cases, blisters can be covered with a heavy gauze pad to reduce pain. It is generally recommended to leave blisters unbroken. When a blister breaks, the area should be cleaned and treated as an open wound. Some blisters tend to be more serious than others. For example, blisters in the palm of a hand or on the sole of a foot should be look at by a doctor.

HAND INJURIES

Hand injuries are common in the flooring trade. Little cuts are the most frequent complaint. Even the smallest break in the skin should be covered. Serious hand injuries should be elevated. This tends to reduce swelling. You should not try to clean really bad hand injuries. Use a pressure bandage to control bleeding. If the cut is on the palm of a hand, a roll of gauze can be squeezed by the victim to slow the flow of blood. Pressure should stop the bleeding, but if it doesn't, seek medical assistance. As with all injuries, use common sense on whether or not professional attention is needed after first aid is applied.

SHOCK

Shock is a condition that can be life threatening even when the injury responsible for a person going into shock is not otherwise fatal. We are talking about traumatic shock, not electrical shock. Many factors can lead to a person going into shock. A serious injury is a common cause, but many other causes exist. There are certain signs of shock which you can look for.

If a person's skin turns pale or blue and is cold to the touch, it's a likely sign of shock. Skin that becomes moist and clammy can indicate shock is present. A general weakness is also a sign of shock. When a person is going into shock, the individual's pulse is likely to exceed 100 beats per minute. Breathing is usually increased, but it may be shallow, deep, or irregular. Chest injuries usually result in shallow breathing. Victims who have lost blood may be thrashing about as they enter into shock. Vomiting and nausea can also signal shock.

As a person slips into deeper shock, the individual may become unresponsive. Look at the eyes, they may be widely dilated. Blood pressure can drop, and in time, the victim will lose consciousness. Body temperature will fall, and death will be likely if treatment is not rendered.

There are three main goals when treating someone for shock. Get the person's blood circulating well. Make sure an adequate supply of oxygen is available to the individual, and maintain the person's body temperature.

When you have to treat a person for shock, you should keep the victim lying down. Cover the individual so that the loss of body heat will be minimal. Get medical help as soon as possible. The reason it's best to keep a person lying down is that the individual's blood should circulate better. Remember, if you suspect back or neck injuries, don't move the person.

People who are unconscious should be placed on one side so that fluids will run out of the mouth and nose. It's also important to make sure that air passages are open. A person with a head injury may be laid out flat or propped up, but the head should not be lower than the rest of the body. It is sometimes advantageous to elevate a person's feet when they are in shock. However is there is any difficulty in breathing or if pain increases when the feet are raised, lower them.

Body temperature is a big concern with shock patients. You want to overcome or avoid chilling. However, don't attempt to add additional heat to the surface of the person's body with artificial means. This can be damaging. Use only blankets, clothes, and other similar items to regain and maintain body temperature.

Avoid the temptation to offer the victim fluids, unless medical care is not going to be available for a long time. Avoid fluids completely if the person is unconscious or is subject to vomiting. Under most job-site conditions, fluids should not be administered.

Checklist of Shock Symptoms

✔ Skin that is pale, blue, or cold to the touch

✔ Skin that is moist and clammy

✔ General weakness

✔ Pulse rate in excess of 100 beats per minute

✔ Increased breathing

✔ Shallow breathing

✔ Thrashing

✔ Vomiting and nausea

✔ Unresponsive action

✔ Widely dilated eyes

✔ A drop in blood pressure

BURNS

Burns are not real common among flooring installers, but they can occur in the workplace. There are three types of burns that you may have to deal with. First-degree burns are the least serious. These burns typically come from overexposure to the sun, which construction workers often suffer from, quick contact with a hot object, like the tip of a torch, and scalding water, which could be the case when working with a boiler or water heater.

Second-degree burns are more serious. They can come from a deep sunburn or from contact with hot liquids and flames. A person who is affected by a second-degree burn may have a red or mottled appearance, blisters, and a wet appearance of the skin within the burn area. This wet look is due to a loss of plasma through the damaged layers of skin.

Third-degree burns are the most serious. They can be caused by contact with open flames, hot objects, or immersion in very hot water. Electrical injuries can also result in third-degree burns. This type of burn can look similar to a second-degree burn, but the difference will be the loss of all layers of skin.

Treatment

Treatment for most job-related burns can be administered on the job site and will not require hospitalization. First-degree burns should be washed with or submerged in cold water. A dry dressing can be applied if necessary. These burns are not too serious. Eliminating pain is the primary goal with first-degree burns.

Second-degree burns should be immersed in cold (but not ice) water. The soaking should continue for at least one hour and up to two hours. After soaking, the wound should be layered with clean cloths that have been dipped in ice water and wrung out. Then the wound should be dried by blotting, not rubbing. A dry, sterile gauze should then be applied. Don't break open any blisters. It is also not advisable to use ointments and sprays on severe burns. Burned arms and legs should be elevated, and medical attention should be acquired.

Bad burns, the third-degree type, need quick medical attention. First, don't remove a burn victim's clothing, skin might come off with it. A thick, sterile dressing can be applied to the burn area. Personally, I would avoid this if possible. A dressing might stick to the mutilated skin and cause additional skin loss when the dressing is removed. When hands are burned, keep them elevated above the victim's heart. The same goes for feet and legs. You should not soak a third-degree burn in cold water, it could induce more shock symptoms. Don't use ointments, sprays, or other types of treatments. Get the burn victim to competent medical care as soon as possible.

HEAT RELATED PROBLEMS

Heat related problems can include heat stroke and heat exhaustion. Cramps are also possible when working in hot weather. There are people who don't consider heat stroke to be serious. They are wrong. Heat stroke can be life threatening. People affected by heat stroke can develop body temperatures in excess of 106 degrees F. Their skin is likely to be hot, red, and dry. You might think sweating would take place, but it doesn't. Pulse is rapid and strong, and victims can sink into an unconscious state.

If you are dealing with heat stroke, you need to lower the person's body temperature quickly. There is a risk, however, of cooling the body too quickly once the victim's temperature is below 102 degrees F. You can lower body temperature with rubbing alcohol, cold packs, cold water on clothes or in a bathtub of cold water. Avoid the use of ice in the cooling process. Fans and air conditioned space can be used to achieve your cooling goals. Get the body temperature down to at least 102 degrees and then go for medical help.

Cramps

Cramps are not uncommon among workers during hot spells. A simple massage can be all it takes to cure this problem. Salt water solutions are another way to control cramps. Mix one teaspoonful of salt per glass of water and have the victim drink half a glass about every 15 minutes.

Exhaustion

Heat exhaustion is more common than heat stroke. A person affected by heat exhaustion is likely to maintain a fairly normal body temperature. But, the person's skin may be pale and clammy. Sweating may be very noticeable, and the individual will probably complain of being tired and weak. Headaches, cramps, and nausea may accompany the symptoms. In some cases, fainting might occur.

The salt water treatment described for cramps will normally work with heat exhaustion. Victims should lie down and elevate their feet about a foot off the floor or bed. Clothing should be loosened, and cool, wet cloths can be used to add comfort. If vomiting occurs, get the person to a hospital for intravenous fluids.

We could continue talking about first aid for a much longer time. However, the help I can give you here for medical procedures is limited. You owe it to yourself, your family, and the people you work with to learn first aid techniques. This can be done best by attending formal classes in your area. Most towns and cities offer first aid classes on a regular basis. I strongly suggest that you enroll in one. Until you have some hands-on experience in a classroom and gain the depth of knowledge needed, you are not prepared for emergencies. Don't get caught short. Prepare now for the emergency that might never happen.

14

TOOLS OF THE TRADE

THE BASICS

In the previous chapters, we've discussed the basic tools required to perform professional flooring installations. Good quality tools are an investment in your professional future, and skimping on quality can result in minor inconveniences that can plague you every day, and on every installation.

Inexpensive tearout tools will bend under steady use and become difficult to work with. Cheap drill bits get dull quickly, and create unnecessary frustration. Poorly designed saws also become dull, and can be awkward to work with. The handles of some low end trowels are spot welded to the blades. Under constant use, these welds can break, which will make the tool completely useless.

Professional grade tools are built to perform at a high level, and will give you the confidence you need to provide your clientele with consistent top notch craftsmanship. The best advice I have for you is to purchase the best quality tools you can afford. Good tools will quickly become a trusted part of the value you bring to your job.

Learn From the Pros

Pay particular attention to the tools that other professionals use during flooring installations. You'll learn a great deal by watching good craftsmen at work, and you'll learn how and why many of the tools that they use help to save time and provide excellent results. Don't be afraid to ask questions about the tools you see in action. Most professionals are proud of their tools and their expertise in using them. Prefacing your questions with a compliment about a pro's craftsmanship will almost always trigger a positive response.

TOOLS OF THE TRADE

In this section of the book, we've provided illustrations and descriptions of a wide variety of professional flooring installation tools. Effective installation of ceramic tile, linoleum, vinyl tile, and hardwood flooring can all benefit from many of the specialty tools shown here, and I encourage you to refer to this section as your expertise increases. you'll find that many of these tools are designed to perform exactly the task at hand, and will save you a great deal of labor and frustration. Believe me, a solid set of tools will pay for itself many times over during the course of your flooring career.

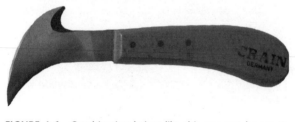

FIGURE A-1 Combination knives like this one can be use to cut both carpet and sheet goods. *(Courtesy of Crain Tools)*

FIGURE A-2 The basic utility knife is an invaluable carpeting and generally floor installation tool. This model features a thumb knob that doesn't require the use of a screwdriver for blade changes. *(Courtesy of Crain Tools)*

FIGURE A-3 Carpet knives that feature a rear loading blade magazine for replacement blades. *(Courtesy of Crain Tools)*

FIGURE A-4 A carpet tool for cutting holes in damaged carpet, and making patches from replacement carpet. *(Courtesy of Crain Tools)*

FIGURE A-5 A loop pile cutting tool, designed to cut from the backing of loop pile carpeting. *(Courtesy of Crain Tools)*

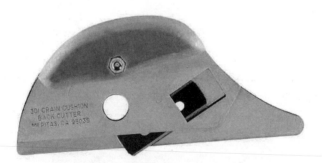

FIGURE A-6 This carpet cutting tool is designed to cut foam and sponge-backed carpet. *(Courtesy of Crain Tools)*

FIGURE A-7 This loop pile cutter is designed to cut low-loop commercial carpet on jute backings. *(Courtesy of Crain Tools)*

FIGURE A-8 The carpet top cutter is designed for accurate seam preparation from the top face of carpeting by overlapping the two edges of carpet, inserting the tool, and then pushing it forward through the length of the seam. The cuts create perfectly aligned edges. *(Courtesy of Crain Tools)*

FIGURE A-9 The thimble palm helps make sewing carpeting easier. Thimbles for sewing are also available in finger thimble models. *(Courtesy of Crain Tools)*

FIGURE A-10 The row separator is a double ended tool used to spread carpet tufts, with one end used for cut piles and the other end for loop pile. This effectively spreads rows for less shearing of nap. *(Courtesy of Crain Tools)*

FIGURE A-11 A typical carpet sewing needle. These needles come in 2⅛, 3, and 4 inch sizes. *(Courtesy of Crain Tools)*

FIGURE A-12 Stretcher legs are designed to permit the stretching of carpeting in long hallways when used with the power stretcher. *(Courtesy of Crain Tools)*

FIGURE A-13 Power stretcher for stretching carpeting. *(Courtesy of Crain Tools)*

FIGURE A-14 Nap shears for cutting carpeting. The duckbill shape prevents gouging. *(Courtesy of Crain Tools)*

FIGURE A-15 Carpet tucking trimmer that rolls the trimmed carpet edge down into the gully at the edges of the room. *(Courtesy of Crain Tools)*

FIGURE A-16 This trimming tool is built to trim both carpet and vinyl to fit walls when used with blades designed for each use. This trimmer can be used to trim accurately under toe spaces. *(Courtesy of Crain Tools)*

FIGURE A-17 This trimmer is designed to trim the proper overlap for use with "Z" bars. This tool cuts straight or contour edges to fit quarry tiles or flat installations. *(Courtesy of Crain Tools)*

FIGURE A-18 Floor scraper is designed to remove vinyl tile and ceramic tile, and built for use from a standing position. *(Courtesy of Crain Tools)*

FIGURE A-19 The inertial scraper is designed with a weighted metal handle to provide inertia for tough vinyl and ceramic tile removal jobs. *(Courtesy of Crain Tools)*

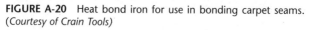

FIGURE A-20 Heat bond iron for use in bonding carpet seams. *(Courtesy of Crain Tools)*

FIGURE A-21 The stripping machine is designed to strip vinyl and ceramic tile with the aid of an electrically powered oscillating blade. *(Courtesy of Crain Tools)*

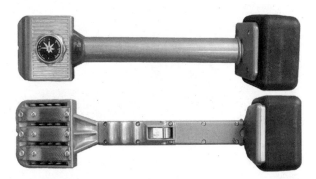

FIGURE A-22 Knee-kicker for stretching carpet. *(Courtesy of Crain Tools)*

FIGURE A-23 This installers tool kit is designed to incorporate the basic tools that are needed for carpet installation by the professional. (*Courtesy of Crain Tools*)

FIGURE A-24 Strap clamp designed to clamp and hold laminate planks across widths up to 18 feet. (*Courtesy of Crain Tools*)

FIGURE A-25   Variable wall spacer is used to space flooring planks from the wall, or to clamp and press butt joints. (*Courtesy of Crain Tools*)

FIGURE A-26   Floor vise is lever-based suction cup tool for the attachment to the faces of laminate planks for hook-ups to clamps, allowing the installer to clamp in any direction, or from anywhere in the room. (*Courtesy of Crain Tools*)

FIGURE A-27 The pull bar is used to drive glued planks tightly together at the tongue and groove joints. (*Courtesy of Crain Tools*)

FIGURE A-28 Plank removal tool is used to remove laminate planks that require replacement. The plank to be replaced is cut into pieces, which are removed with the prying action of the tool. (*Courtesy of Crain Tools*)

FIGURE A-29 Dual air sled appliance moving system. When the footplates are slid under the load and the blower is turned on, the load is lifted. The load can be moved with the touch of a fingertip. (*Courtesy of Crain Tools*)

FIGURE A-30 The carpet cart is used for supporting and carrying full or partial rolls. (*Courtesy of Crain Tools*)

LEEDS COLLEGE OF BUILDING
LIBRARY

FIGURE A-31a Dollies designed for moving flooring material. (*Courtesy of Crain Tools*)

FIGURE A-31b Dollies designed for moving flooring material. (*Courtesy of Crain Tools*)

FIGURE A-32 Carpet seaming roller designed to pull seams together while mixing the carpet fibers for less visible seams. (*Courtesy of Crain Tools*)

FIGURE A-33 Carpet seaming roller with ribbed rollers, which makes it especially good for saxony, plush, and velvet carpets, because it will not snag the yarn. (*Courtesy of Crain Tools*)

FIGURE A-34 Seam squeezers force the edges of cushion back carpet together, producing an improved seam. (*Courtesy of Crain Tools*)

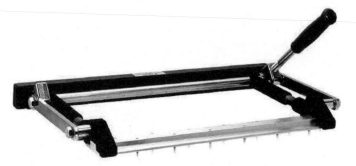

FIGURE A-35 Stairway stretcher is designed to work from the top of the stairway down, allowing gravity to work with the installer. (*Courtesy of Crain Tools*)

FIGURE A-36   The seam repair stretcher is designed to pull together open carpet seams for repair with latex; tape, or an iron. (*Courtesy of Crain Tools*)

FIGURE A-37   Hand stretcher is used in cramped areas near the wall. (*Courtesy of Crain Tools*)

FIGURE A-38   Narrow stair tool. (*Courtesy of Crain Tools*)

FIGURE A-39 Wide cushion stair tool. (*Courtesy of Crain Tools*)

FIGURE A-40 Bent stair tool. (*Courtesy of Crain Tools*)

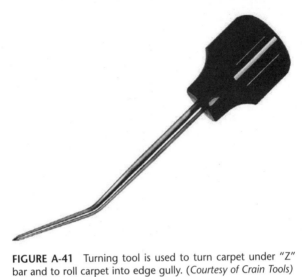

FIGURE A-41 Turning tool is used to turn carpet under "Z" bar and to roll carpet into edge gully. (*Courtesy of Crain Tools*)

FIGURE A-42 Carpet awl with a high-carbon steel pointer that extends through the handle for hammering. (*Courtesy of Crain Tools*)

FIGURE A-43 Push cutter designed to slice glued down carpet into strips, or for rough-cutting carpet off the roll. (*Courtesy of Crain Tools*)

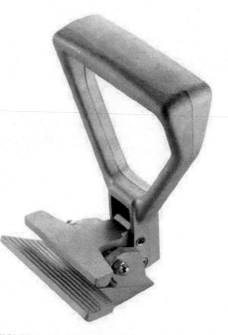

FIGURE A-44 Tear-out tool designed to clamp tighter with pulling pressure. (*Courtesy of Crain Tools*)

FIGURE A-45 Carpet clamp used for tear-out jobs. (*Courtesy of Crain Tools)*

FIGURE A-46 Toe-kick saw. (*Courtesy of Crain Tools)*

FIGURE A-47 Toe-kick saw in use. (*Courtesy of Crain Tools*)

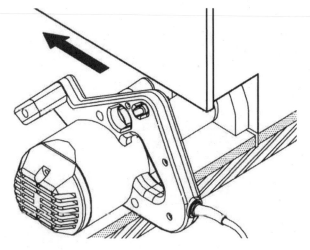

FIGURE A-48 Toe-kick saw, direction of use. (*Courtesy of Crain Tools*)

FIGURE A-49 Magnetic tack holder reaches into the carpet pile where fingers won't go, which makes this a good tool for stair work. (*Courtesy of Crain Tools*)

FIGURE A-50 Door pin tool for the romoval of door pins without damage. (*Courtesy of Crain Tools*)

FIGURE A-51 Multi-undercut saw undercuts walls, door jambs, under toe-spaces and inside corners. (*Courtesy of Crain Tools*)

FIGURE A-52 Heavy-duty undercut saw can undercut most doors in place. Used with a masonry blade, this saw will also undercut hearths. (*Courtesy of Crain Tools*)

FIGURE A-53   Jamb saw used for undercutting doors, door jambs, and masonry walls. (*Courtesy of Crain Tools*)

FIGURE A-54a   Jamb saw in use for undercutting wood and masonry. (*Courtesy of Crain Tools*)

FIGURE A-54b Jamb saw in use for undercutting wood and masonry. (*Courtesy of Crain Tools*)

FIGURE A-54c Jamb saw in use for undercutting wood and masonry. (*Courtesy of Crain Tools*)

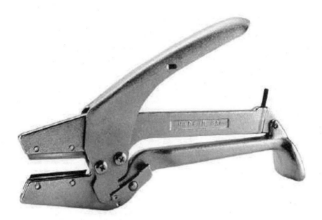

FIGURE A-55 Strip cutter is designed with two blades to cut from the top and the bottom. (*Courtesy of Crain Tools*)

FIGURE A-56 Extension wall roller. (*Courtesy of Crain Tools*)

FIGURE A-57 Solid steel vinyl roller with roller sections that float independently to compensate for sub-floor irregularities. (*Courtesy of Crain Tools*)

FIGURE A-58 Vinyl seam roller. (*Courtesy of Crain Tools*)

FIGURE A-59 Vinyl tile scribers. (*Courtesy of Crain Tools*)

FIGURE A-60 Outside corner scriber for scribing outside cove corners. (*Courtesy of Crain Tools*)

FIGURE A-61 Bar scriber telescopes to make scribes from 1 inch to 23 inches. The roller tip guides against the wall. (*Courtesy of Crain Tools*)

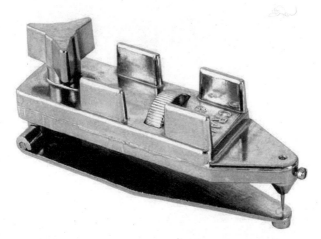

FIGURE A-62 Hinge scriber for use with linoleum, sheet vinyl, and cushion flooring materials. (*Courtesy of Crain Tools*)

FIGURE A-63 Pin vise for scoring tile and linoleum pattern layouts. (*Courtesy of Crain Tools*)

FIGURE A-64 Vinyl tile cutter. (*Courtesy of Crain Tools*)

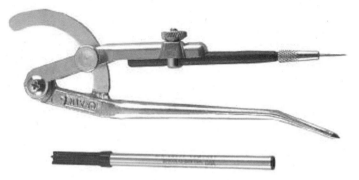

FIGURE A-65   Divider featuring a bent leg and a removable pin vise with a knurled tightening cap. (*Courtesy of Crain Tools*)

FIGURE A-66   Handle and die-set combination for vinyl and laminate cap metal, designed to create corners without a miter box. (*Courtesy of Crain Tools*)

GLOSSARY

B

Backing yarn   Anchors pile fibers in a piece of carpet.

Baseboard   A board placed against the wall around a room next to the floor, properly finishing the area between the floor and wall.

Base molding   Trim that sits against the baseboard and rests on the floor.

Bedding   A layer of mortar into which brick or stone is set.

Bisque   Special material applied to ceramic tile prior to firing to obtain a glazed finish.

Blind nailing   Nail head not visible on the face of the work, usually placed in tongue of matched board.

Bowing   A twisting out of shape caused by unequal stress.

Brace   Diagonally framed member used to hold a floor in place.

Bridging   Diagonal metal or wood braces installed between joists to spread the load and prevent twisting.

C

Cap molding   Trim applied to top of base molding.

Caulking   Elastic adhesive or plastic polymer used to seal joints.

Ceramic tile   Mixture of baked clays made into a variety of shapes and colors.

Cement board   Placed under ceramic tile, used to maintain moisture protection.

Cement mortar   Mixture of cement, sand, and water used as a bonding agent between bricks and stones.

Cleat   Length of wood secured to surface to maintain an object in place.

Concrete   Cement, stone, gravel, and water mixture causing cement to set and bind.

Cork tile   A tile made of cork, sandwiched between vinyl backing and clear vinyl top layer.

Cut-pile   A carpet surface that reveals cut end of each end of yarn.

D

Direct nailing   To nail perpendicular to initial surface or junction of pieces joined. Also known as *face nailing*.

Drop match   Indicates carpet pattern is situated at a diagonal across a carpet width.

E

Edge nailing   Nailing into the edge of a board.

Embossing leveler   Mortar-like substance which can be used as an underlayment between well adhered resilient or ceramic tile.

End nailing   Nailing into the end of a board.

Expansion joint   Joints between an area designed to expand and contract.

F

Filler   Putty or other pasty materials used to fill nail holes prior to painting sanding or staining.

Finish   A final surface applied to a floor.

Floating   A process used after screeding to provide a smoother surface to a concrete floor.

Floating floor   Boards of a floor are glued to each other, but not to sub-floor. Held in place by their own weight.

Floor leveler   Compound used to fill in dips and low spots in a plywood floor.

G

Glaze   Hard, shiny surface fired on the surface of ceramic tiles.

Glue-down floor   Adhesives used on a sub-floor to attach flooring.

Granite   A natural stone that occurs in many colors.

Grout   A cement-based material used to fill in spaces between tile.

Gypsum board   Interior floor or wall covering material applied in large sheets or panels.

I

Isolation membrane   Used to protect ceramic tile installations from movement that may occur on cracked concrete floors.

J

Joint   Space between adjacent surfaces that are held together by nails, glue, cement, mortar, or other materials.

Joist   Long pieces of lumber used to support floor boards.

K

Kiln-dried   Wood dried with the use of artificial heat.

L

Landing   Platform between flight of stairs.

Lap   A joint of two pieces lapping over each other.

Latex-modified grout adhesive   A blend of acrylic and latex decreasing water absorption, but increasing strength and improving color retention.

Level-loop   Carpet pile formed with loops of yarn cut to a uniform height.

Limestone   Calcite from shells, coral, and other debris.

Loop-pile   Carpet constructed of a surface of continuous loops.

M

Marble   Any limestone or corbonate, granular to compact in texture, that will take a polish.

Masonary   Stone, brick, concrete, hollow tile, or concrete block blended together by mortar.

Mastic   Any pasty material used as a cement to set tile.

Molding base   Molding on top of a baseboard.

Mortar   A mixture of cement, lime, sand, and water often used as a setting bed for tile.

Mosaic   Setting of small tile in a pattern or picture.

Mud   Slang for spackle or drywall compound used to seal joints and hide nail heads.

N

Nap   Direction that the pile of a carpet tends to lie.

Natural finish   Transparent finish which does not seriously alter the original color or grain of a natural wood.

P

Parquet   Floor with an inlaid design.

Penny   Term applied to a nail, originally indicating price per one hundred. Now used as a measured length of a nail, abbreviated by the letter d.

Perm   A measure of water vapor movement through a material.

Pile density   Number of tufts per unit square used in a carpet.

Plank   Wide piece of sawed timber, usually ½ inch to 4 inches thick and 6 inches or more wide.

Plush   All loops of carpet cut to a uniform height.

Plywood   Piece of wood made of three or more layers of veneer, joined with glue.

Polyethylene   Heavy-gauge plastic used as a vapor barrier.

Poultice   A powder or cloth placed on a stain, which is designed to remove stain by absorption.

Q

Quarry   Name given to place where stone is removed from the ground.

Quarry tile   An unglazed ceramic floor tile.

R

Radiant heating   Hot water flows through tubing in a sub-floor, warming the floor, in turn heating the room.

Reducer   Wood transitions used between hardwood flooring and an adjacent floor of a lower height. One edge is grooved while the other is bull-nosed.

Refinish   Process of returning a surface to its original finish.

Ring shank nail   Nail with ridges forming around the shank providing better withdrawal resistance.

Rise   Vertical height or stop of flight of stairs.

Riser   Vertical board between two treads of a flight of stairs.

Run   For stairs, the net front-to-back width of a step or horizontal distance covered by a flight of stairs.

S

Sanded grout   Cement based grout containing sand particles.

Screed   Small strip of wood, usually the same thickness as a plaster coat.

Sealer   Finishing material, either pigmented or clear, applied directly over an uncoated piece of wood for the purpose of sealing.

Setting   Act of installing stone or tile.

Set match   Indicates pattern is in a straight side-by-side match across width of carpet.

Slab   A concrete floor poured on the ground.

Slate   Stone metamorphosed from shale.

Slip tongue   Spline used to connect two adjacent boards that have grooves facing each other.

Stripping   Process of removing waxes or other coatings from the surface of a floor.

Sub-floor   Rough floor, concrete slab or wood over joists, onto which final floor is laid.

T

Tack-less strip   Thin board ⅛ inches by 1½ inches wide with two or three rows of rust resistant metal pins pointing upward at a 60 degree angle. Used to secure edges of carpet after stretching.

Terrazzo   A marble, chip, and cement mixture.

Threshold   Strip of wood or metal with beveled edges placed over the edges of a finished floor between an interior or exterior door.

Tongue and groove   Boards that join on edge with a groove on one unit and a corresponding tongue on the other to interlock. Also known as *dressed and matched*.

Tread   Horizontal board in a stairway.

Troweling   Process after floating to provide a smooth concrete surface.

U

Underlayment   Any paper or felt composition used under flexible flooring, such as vinyl or carpet, to provide a smooth base over which to lay such materials.

V

Vapor barrier   Material used to retard movement of water vapor and prevent condensation.

Varnish   Dry oil or oil and resin suitable for spreading on surfaces to form a continuous, transparent coating.

W

Warping   Twisting out of shape due to unequal stresses or bowed due to relatively weak restraining forces on the opposite side.

Waterproof membrane   Stops moisture from seeping up or down through flooring.

Water spotting   Spot produced on a floor surface due to sprinkling or dropping of water

INDEX

NOTES

NOTES

NOTES

NOTES